Military Resilience in Low-Intensity Conflict

Military Resilience in Low-Intensity Conflict

A Comparative Study of New Directions Worldwide

Rachel Suissa

LEXINGTON BOOKS
Lanham • Boulder • New York • Toronto • Plymouth, UK

Published by Lexington Books
A wholly owned subsidiary of The Rowman & Littlefield Publishing Group, Inc.
4501 Forbes Boulevard, Suite 200, Lanham, Maryland 20706
www.rowman.com

Estover Road, Plymouth PL6 7PY, United Kingdom

British Library Cataloguing in Publication Information Available

Library of Congress Cataloging-in-Publication Data

Suissa, Rachel, 1954-

Military Resilience in Low Intensity Conflict: A Comparative Study of New Directions Worldwide / by Rachel Suissa.

p. cm.

Includes bibliographical references.

ISBN 978-0-7391-2832-9 (cloth : alk. paper) — ISBN 978-0-7391-7287-2 (electronic)

1. Low-intensity conflicts (Military science)—Case studies. 2. Asymmetric warfare—Case studies. 3. Resilience (Personality trait) I. Title.

U240.S97 2012

355.02'15—dc23 **2011039763**

Printed in the United States of America

∞™ The paper used in this publication meets the minimum requirements of American National Standard for Information Sciences—Permanence of Paper for Printed Library Materials, ANSI/NISO Z39.48-1992.

To Raphael, the most devoted father.

Contents

Introduction

Military resilience is a concept that includes structural-military aspects of force building and technology among others, functional aspects of the warfare system, and mental-social aspects, such as morale, endurance and military-society relations. In the professional literature, all of these are derived from the national resilience of the state. Various facets of military resilience have been discussed in the literature, such as relations between society and the military in prolonged conflicts (Gal, 2002; Levy, 2004 and others), dilemmas of national security versus democracy, and the implications of these on national-military resilience (Yaniv, 1993; Ben Dor, 1993, 1998; Neuberger, 1996; Dror, 1999; Al-Haj and Ben-Eliezer 2003 and others), and conscientious objection, motivation, reserve service. (Shnitzer, 1997; Gal, 1986; Lin, 1997). Regarding the Grand Strategic level, the military resilience of the variety of clusters has been discussed within the framework of military doctrine and has been derived from the connection between goals and objectives and the means to achieve these goals on the level of conflicts between states (Posen, 1984) and of asymmetrical conflicts (Burke, 2003).

As the problem of resilience is interdisciplinary, it is discussed using a variety of research approaches from the areas of sociology, psychology, psychiatry, and organization theory (Bartone 1999; Bartone 2000; Gal 1987). These approaches have held dominant contributions to the study of military resilience in the terminological, theoretical and practical terms. Yet, the concept of military resilience has been excessively "stretched" in psychological and psychiatric paradigms. This has limited the range of possible variables that affect resilience, and became difficult to formulate a specific theoretical framework for understanding the problems that would include an inherent conceptual system for the complexity of the military arena with all its structural and functional aspects. An illustration of "excessive stretching" can be found in the various approaches dealing with the problem of resilience using the terms *psychological stress*, *cognitive stress*, *trauma*, and *post-traumatic stress*. These concepts are applied in connection with performance with coping,[1] with mediation resources between pressure and coping,[2] and with the theory of psychological stress that has succeeded in presenting emotional and cognitive models for coping with stress.[3] The concept of "stress" has been too greatly "stretched" to contribute to an understanding of the dilemma, which requires the identification of additional necessary concepts to extract its full meaning.

This book, on the one hand, extends the range of concepts for understanding the problem of resilience, and on the other, it limits the problem to the military sphere. The central rationale for this research consideration is based on a number of premises:

Not every aspect of the problem of resilience is related to stress, whether personal, psychological or cognitive. All resilience, whether human-biological and/or organizational may be harmed by internalizing insights consciously and rationally, with *no accompanying experience of stress and with no elements of post-trauma, among others.* Then, if these insights are not in line with expectations of any kind, the harm to resilience may be expressed in a deliberate disruption of continued functioning. This disruption is not a result of stress, but rather a rational calculation resulting from the non-congruence between new insights and previous expectations.

Precisely because stress theory is so natural for understanding the resilience of soldiers in every military sphere, while in military reality we are accustomed to discussing this, we must divert the discussion from something that is usual to something that is not. In other words, we should not project the theory of emotional stress, trauma or post-trauma on to the issue of resilience in every situation. A low-intensity conflict still remains "an unusual situation" for many armies, and forces them into a different kind of thinking from what they were used to until considering the problem of resilience and the ways to deal with it.

The army, as a military organization, is at the basis of this book. Thus, an observation of the problem of military resilience takes the human factor into account as central, but it is analyzed through a structural-organizational-collective prism. Through this prism, the use of the theory of stress and trauma to understand military resilience is particularly limited. Therefore, the terminology of the dependent and independent variable concepts proposed for discussion are structural-organizational-collective aspects, as well. The military model, warfare strategy and relations between the political-military ranks embody these kind of aspects, and are proposed as independent variables affecting resilience.

Psychological and psychiatric research for the most part relates to the fighters, for whom any harm to them in warfare management involves their continuing military and civilian functioning, while they no longer usually belong to warfare forces. These soldiers have been harmed by their being part of and acting according to the operating principles derived from the universal duties of armies in general, moving between "killing" and "dying" with everything that is implied therein concerning the individual's mind. Therefore, this book is concerned with broadening the discussion of resilience that is relevant to the general human factor that manages the warfare campaign and whose horizons are beyond what has been known throughout history about the difficulties of the human mind in managing warfare activity.

Military Resilience

Beyond the proposal of a new theory in the sphere of military resilience in a low-intensity conflict, the book also, for the first time, compares the military resilience of states that have been or are now involved in low-intensity conflicts.

In contrast to the reality of classical large-scale warfare, a low-intensity conflict represents a warfare reality that is different in nature, beginning in the general rules of using force, unique operational perceptions in a unique arena and ending with problems of cohesion and motivation particular to this situation. All of these characterize the asymmetry of the nature of the conflict. Moreover, the identification and characterization of a low-intensity conflict demands adherence to its own dynamic. Specifically, a low-intensity conflict can be the continuation of a full-scale war, a separate episode that develops and changes, and eventually in certain circumstances, may develop into full-scale warfare.

The uniqueness of a low-intensity conflict requires a different logic from that of a discussion of dilemmas of resilience during normal periods or involved in a full-scale war (van Creveld, 1991; Hammes, 2004; et al.). This book discusses military resilience in a low-intensity conflict. However, resilience is a different challenge in different situations. A full-scale war places unique demands on the resilience of the military organization, and routine times has their own demands on resilience. Though the latter is not characterized with significant operative loads, it possesses challenges to military resilience.

A low-intensity conflict refers to a situation in which the adversaries use opposing rules of the game while, at the center, an army may lose the abilities that have been developed in light of its usual missions. In other words, the objectives of the warfare and the training framework of regular armies suit the "rules of the game" of a warfare arena between states and regular armies. The goal of the fighting is to defeat the other army decisively. However, to fighters of irregular military organizations, the very concept of a front is not applicable; it is indistinct and is used in confusion with the concept of "home front." These organizations are not interested in resolution at the first stages of warfare, but rather, the opposite: indecisiveness. The objective of the fighting is to maintain their independent existence by adopting methods of warfare in installments, exhausting the enemy, wearying him, eroding his strength, attacking his weak points and accumulating small victories that will tip the scales in his favor. Thus, the constraints of fighting terror, to a great extent, challenge the abilities that emanate from the classic missions of an army.

The author of this book is well aware of the transformation from full-scale warfare to low-intensity warfare. An especially salient contribution to the professional literature has been *The Transformation of War* by Martin van Creveld (1991). Creveld maintains that the era of full-scale war has ended and that humanity is now facing guerilla warfare and irregular military organizations. These wars, low-intensity conflicts, have constituted the main form of warfare since the end of World War II, and have been of decisive significance in terms of victims and in their minimal chances for political achievements. The classic Western warfare perception is not relevant any more to the management of this type of conflict, which is viewed by van Creveld as a non-political conflict, and he expects a transformation to religious and ideological warfare with a propensity of forms of fanaticism. Technologically sophisticated weapons will disappear and will give way to simpler munitions and to the collapse of the international arms industry.

The development of the discussion of this book that is based on empirical analysis of four different conflicts, has indicated that the transformation of warfare is not only in one direction. It is true that, from the end of 1945, we have been witnessing a transformation from conventional warfare to low-intensity conflicts. However, as in every process of transformation, we are also witnessing a rise in the feasibility of full-scale war. Thus, the transformation described is a reversible process. Moreover, this book attempts to point to a transformation from low-intensity conflict to full-scale war in situations in which a terrorist movement has become a guerilla movement and then, an irregular military organization. This returns the rationale for classic military warfare to the security agenda. The term "flexibility" in the various warfare behaviors should be included in a discussion about warfare transformation, as appears, for example, in the discussion at the end of *The Sling and the Stone* by T.X. Hammes (2004) who analyzed the development of conflicts of this type in the context of "conflicts of the fourth generation of the modern era."

As for the idea that we are in an era where full-scale wars have ended and that humanity is now facing mainly non-political, religious, and ideological conflicts: the principle of transformation developed by Van Creveld seems limiting the transformation from fanaticism to pragmatism that radical organizations go trough when institutionalized. This means that the political character becomes a most prominent and relevant aspect within a transformation from fanaticism to pragmatism. Paradoxically, this increases full-scale wars' propensity.

Indeed, war transformation is the issue. Yet, we should avoid the illusion the linear significance this term might suggest in shifting from one war pattern to another as well as from the analytical aspect derived from wars classifications. Not only do war transformations work in different trends, but a state might conduct different level of wars at the same time. This means asymmetrical wars can be transformed to full-scale warfare and vaice versa, and different types of war can take place in one battle scene. Thus, it is not about the conduct of this kind or another of war, but of different war categories at the same time. An illustration of this can be found in the Israeli case for instance. According to its threat perception evaluation, Israel might experience a low-intensity conflict with Hamas parallel to a full scale war with an Arab state like Syria while trying to cope with a non-conventional warfare with Iran for example.

Another point illustrated in the empirical part of this book is the chronology of low-intensity conflicts, from the 1954–1962 French-Algerian, through the 1964–1994 Britain-Ireland war, 1994–2000 Russia-Chechnya war to 2000–2004 Israel-Palestinian war. This chronology implies that war transformation takes place not only from one war to another, but also within the same category of war. Thus, low intensity conflicts have gone through transformation not only because these conflicts differ from each other, but as phenomenon in themselves.

Since civil-military relations has implications on the management of a low-intensity conflict due to the permeable borders between the military and the civilians, this perspective that is most relevant to a discussion of military resilience should be clarified, as this is what directs such a long-lasting conflict.

Beyond the need for improved clarification of terms, the basic assumption is that armies usually have both the tools to deal with problems of resilience and the ability to place these problems in proportion so that they do not damage military resilience. The dilemma begins in particular when this resilience is harmed or is eroded in long-term low-intensity operational activity. The understanding of the essence of this resilience, its characteristics and methods of preventing damage to it may be provided in an analysis of the experiences of states that were or still are conducting low-intensity conflicts. The conflicts discussed are France-Algeria 1954–1962, Great Britain-Northern Ireland 1964–1994, Russia-Chechnya 1994–2000, and Israel–The Palestinian Authority 2000–2004 (the death of Arafat).

The rationale at the basis of the comparison between these conflicts is the similar asymmetry characteristics they share as low-intensity conflicts is that the low-intensity conflicts in these countries are characterized by most of the characteristics of asymmetry. In addition, the independent variables that will be discussed as affecting military resilience (the military model, the warfare strategy and the relations between the political-military echelons) enable a relevant comparative debate that on the one hand, grants validity to the variables, but it enables a determination of their limitations on the other. Therefore, the discussion of these states should supply valuable conclusions, both on particular and on universal levels, and thus, enable the construction of the first theory of its kind in the area.

An interesting fact regarding the cases discussed is the sequence of time beginning from the French-Algerian conflict and continuing up to the Israeli-Palestinian conflict. This chronological sequence provides a research framework for observing the development of a low-intensity conflict as a phenomenon that undergoes changes, involving military, technological, political, judicial, and global facets. The book does not deal with the chronological aspect and its analysis in these terms.

The comparative study that was at the heart of the manuscript speaks about four cases stretching over fifty years, which is to say, different historical periods, in different countries, ranging from a geo-strategic giant like Russia to small Israel. In order to make the comparison between such different cases in time and space the book provides systematic and disciplined set of variables:

- The Military Model variable encompasses three different epochs to allow a flexible comparison to the different countries and periods
- The Warfare Strategy variable reflects maximal components of low-intensity conflicts and allows to trace transformations in warfare
- The Political Military Relations variable aims at basic conventions suitable to all political systems at war.

As for the dependent variable, Military Resilience, the detailed clusters provided allow a true scanning of the military resilience profile. Thus, this research is aware of the vast differences between the countries and the periods under comparison. These differences provide internal as well as external validity to the comparison this book conducts. The book chooses the perspective of low-intensity conflicts as posing similar challenges to different countries in different epochs and not vice versa.

The book deals with a number of theoretical questions, and explains how does the military model affect military resilience; how this is possible to describe this effect on military resilience in a low-intensity conflict. The book argues that the type of military warfare conducted in low-intensity conflicts affect military resilience, and the relations between political and military ranks have dominant effects too on military resilience. Thus, the book offers a new theory regarding military resilience that is beyond the mission, load, stress and so on. The empirical aspects of this theory examines how have France, Britain, Russia and Israel dealt with the problem of military resilience in a low-intensity conflict, and what can be concluded regarding possible solutions to this problem.

Chapter 1

Theoretical Background

This book is a first attempt to attribute the effects of the military model, the conflict warfare strategy, and the relationship between the political-military echelons, on the military resilience in low-intensity conflict. The present study will attempt to direct the significance of these variables to the unique warfare reality that characterizes a low-intensity conflict. Over and above the descriptions provided in professional literature dealing with topic of the research, the literature survey will indicate the dilemmas that are raised regarding a variety of variables and their possible solution.

Military Resilience in Low-intensity Conflict

Beyond its military-social-national context, military resilience is discussed as a multi-disciplinary concept in the interdisciplinary framework of the socio-psychological approach, organizational psychology, and military psychology. The closest discussion to the problem of resilience studied here has been conducted in the areas of organizational and military psychology. In these approaches, the discussion of military resilience has centered on a comprehensive identification of sources of stress that are induced by a military environment in an attempt to derive an understanding of the coping resources of the individual participating in a military mission.[1] The conventional battlefield is perceived as one that includes high physiological demands that affects the individual and represents a threat to the individual's wholeness.

However, this discussion does not provide an understanding of the problem of resilience in the unique sphere of a low-intensity conflict. The level of discussion up to now has focused particularly on the individual (soldier) and/or on the organizational-military environment. Even the meager literature that deals specifically with military resilience in low-intensity conflicts continues to focus on concepts of stress on the combatants.

A research on American soldiers who participated in the peace-keeping mission in Bosnia, as a type of mission characteristic of a "low-intensity conflict" that causes stress to the soldier and his family,[2] points to the stressful characteristics of this type of conflict such as the lack of knowledge of when the mission will be completed, long working hours, among others.

Similar findings appear in the most recent research.[3] The conceptual basis for theories discussing resilience deals with stress and post-trauma paradigm. This is also true of psychological-cognitive theories that have contributed much to proposed treatment strategies designed to prevent stress and to maintain emotional resilience. Janis[4] was one of the first to deal with proposals to prevent stress. He suggested the concept of "battle resilience" (that he later changed to "emotional resilience") and hypothesized that it would be possible to devise training plans that would provide simulation to what would be experienced on the actual battlefield.

This book proposes that understanding the problem resilience should be broadened and that the discussion of military resilience should not be replaced by a discussion of the factors that damage resilience. In other words, the object of the investigation should not be identified by the factors that affect it:

- The resilience of the army represents, first and foremost, its performance value on a functional and mental level, which is derived from the nature of the unique qualities of each military organization.
- The content of resilience must be differentiated clearly for the specific situation, which is, in this case, a low-intensity conflict.

The tensions and pressures on the individual and on the environment, and how the individual copes with these do not supply comprehensive understanding of the problem of resilience in a low-intensity conflict. They are closer to a description of the situation and an identification of the environmental characteristics. However, it should be remembered that, in every discussion of resilience of a military organization, and in regard to any organism, in general, there is an implicit demand or requirement for continued proper and stable functioning. This is the essence of resilience against the backdrop of environmental and temporal challenges. This "essence" must be expressed as a salient component in its definition, as well.

Indeed, definitions in this direction have existed and have been discussed extensively.[5] These are directly relevant to the problem of military resilience, as, for example, "the measure of resilience is in the ability to maintain the continuous functioning of a combat framework" or the fact that "mental maintenance of an army involved in a low-intensity conflict is the given significance by officers," among others. These definitions lack due to the reduction of the term "resilience" and remain elusive.

The most competing hypotheses and theory this book wishes to refer to has been discussed within a "synonym" to "military resilience" that Bartone refers to as "hardiness."[6] He links this term with the integral part of leaders' influence, and argues that highly effective leaders can increase hardy, resilient responses to stressful circumstances within their units. He focuses on modern military operations and identifies five primary psychological stress dimensions in modern military operations: isolation, ambiguity, powerlessness, boredom, and danger.[7] These stress dimensions hold a significant theoretical as well as methodological contribution to the study of military resilience. Yet, they all focus on the individual personality without a proper integration of organizational, structural and functional aspects. Since they are categorized as "stress" dimensions, they do not provide a solution to the limitations raised above concerning the "stretching" of this concept. Moreover, these stress dimensions can themselves be triggers or by-products of more relevant dimensions. If we broaden the concept of military resilience beyond "stress" or "hardiness" we might allow a wide spectrum range for military resilience dimensions.

Defining Military Resilience in a Low-intensity Conflict

It is impossible to provide a definition of military resilience without making a clear distinction between an army in a period of normalcy and in comprehensive warfare, and an army in a period of continuing low-intensity conflict. In other words, the definition of resilience is dependent upon the situation, and assurance of military resilience in periods of normalcy or in comprehensive war should not be seen as assurance of resilience in a low-intensity conflict. The nominal definition of military resilience in a low-intensity conflict is as follows: ***"Military resilience in a low-intensity conflict is a measure of stability in the effective implementation of missions over time while maintaining the Building of Force, Unit Resilience, the Military Framework, Significance in the Realm of Military Spirit, Significance in the Realm of Operational Fitness."***

Regarding the "***measure of stability for effective implementation of missions***": It does not only mean effectiveness in the evaluation of results, but it also demands a basis for consensus between the political and military echelons, which does not stray from the goals of the combat, as they have been determined by the political echelon.

The term "***missions***" means that general missions must be included, that is, missions with which the army is experienced, such as building force, alongside specific missions in conducting a low-intensity conflict.

The definition attempts to point out most of the expressions of damage to resilience, that is, "***a blow to the building of force, to unit resilience, to the military framework, to the spirit of the military corps and professional fitness.***" Military resilience in a low-intensity conflict will be tested in five clusters:

Building of Force

1. The extent of fitness, training, and the quantity of units in training.
2. The capability of long-term preparation.
3. The ability to utilize resources and strength building abilities by the units.
4. The framework of the training.
5. Utilization of professional expertise.

Unit Resilience

1. The ability to maintain the organic framework, in a way that either harms or maintains the identification and the internal cohesion.
2. Coping with the burden of activity that leads to fatigue.
3. Organization of unit belonging and identity.

Military Framework

1. The ability to carry out the orders and procedures of a professional army.
2. Maintaining channels of internal and external organization (orders, procedures, messages, intentions).
3. Maintaining learning systems for encoding and distributing internal and external organizational information.
4. Managing tension between what is important and what is urgent in the realm of operational routine.

Significance in the Realm of Military Spirit

1. Expressions of fear and anxiety, exhaustion etc.
2. Questions about the significance of the mission.
3. Assembling candidates for officer positions.

Significance in the Realm of Operational Fitness

1. Activating technological methods, which create dilemmas in the realm of maintaining service norms and fitness, that is, the use of technological methods when they are not congruent with the service norms, and the readiness to act accordingly.
2. The number of accidents and operational mishaps.
3. Defining required fitness that derives from the tension that exists in preparing for a high intensity conflict in relation to a low-intensity conflict.
4. The large gap in assembling fighting support and in the readiness to sign up for long-term military service.

A Methodological Dilemma

Determining the Level of Damage to Military Resilience should relate to its dimensions:

Great Damage—Identifying damage to one element or more in at least two clusters.

Minor Damage—Identifying damage to one or more elements in one cluster.

The reason in these levels is in the significance of specific damage in comparison to systemic damage to military resilience. The five clusters have a logical and practical connection with one another. Specific damage in one cluster may cause damage in another cluster, and damage to more than one cluster makes systemic damage inevitable. It can be assumed that not every element in each cluster has the same importance. However, a situation is possible in which damage to all of the elements in one cluster will be considered less severe than damage to two elements in different clusters whose importance is considered relatively less critical. Although this book does not deal with a calculation of the relative importance of each element in relation to other elements in the same cluster, this problem can be solved logically, and practical support for the solution can be provided. The solution is based on the fact that, despite the differences in levels of importance among the various elements of each cluster; the fact that these elements belong to the same cluster creates reciprocal relations between them. This means that damage to one element, important or unimportant, may damage another element in the same cluster. On the other hand, this damage may be "compensated" by another element in the same cluster. Put differently, a specific damage to one or more elements in a cluster may become more moderate as a result of the reciprocal relations that exist among the elements and perhaps may even be reversed.

In contrast, in a situation in which the damage spreads and influences other elements of the cluster, systemic damage may be foreseen in the absence of a mechanism for correction and/or compensation among the various clusters in a situation of damage to more than one of them.

Due to the importance of grading the damage to resilience and the need to grant validity to this classification, questions should be asked that require additional substantiation to determine the severity of the damage to resilience. In other words, logically, there is still a need to explain why the reciprocal relations between the elements in the same cluster are characterized by spreading damage or compensation, and why the reciprocal relations between the various clusters are characterized only by spreading damage. Furthermore, it would appear that the present explanation predicts two contradictory results regarding damage to an element of the cluster (spreading damage/compensation), and does not explain when each of these will occur and/or that will occur with greater probability. Is there anything in these questions that will destabilize the explanation of the proposed separation between light damage and serious damage?

If we do not separate logical necessity from the content of the clusters, that is, if the question of the logic in the division between minor and grave damage does not ignore the content of the clusters themselves, it is possible to solve these problems. In other words, logical considerations will not be disconnected from their objects but will be internal logical considerations, dependent on content that does not only require relations between clusters and their components but that also takes into consideration their own content and concerns. To what are we referring?

A. *A similarity exists between the content and the components of each cluster in contrast to the differences between various clusters.* The title of each cluster alone is enough to testify to the common denominator between its components, on the one hand, and to the difference between each cluster and the others, on the other. Therefore, if the logical considerations remain within the framework of the data, their solution will be found using this data.

The reciprocal relations among the components of each cluster should be understood against the background of their common denominator. It is this common denominator that makes it possible to claim that these relations can be characterized either by spreading damage or by compensation. Moreover, the common denominator creates a kind of system in which the components of each cluster constitute its parts. If we remember that the research focuses on the human factor of military resilience, an aspiration for homeostasis may be attributed to this system, that is, the aspiration for equilibrium. Thus, not only is it possible to claim that these reciprocal relations may be characterized by spreading damage or by compensation, but that their existence as components of the cluster, as a system aspiring to equilibrium, shows that there is no contradiction in the predication of two opposing results in the case of damage to the components of one cluster or another.

This may be illustrated, for example in clusters A and B. In cluster A the common denominator of the five components is the building of force. The connection between training-planning-utilization of resources and the ability-framework for professional training-expertise makes possible:

1. Direct and immediate reciprocal relations between these components.
2. Spreading damage or compensation. How is this possible? Damage to the training framework (A4) may cause damage to the scope of training sessions, the training itself, the number of units participating (A1), and vice versa. However, this damage can be compensated for in the more intensive utilization of resources, and the ability to build existing force, as an expression of the additional effort that has become required due to the damage. If utilization of professional expertise is maintained, a kind of compensation of quality over quantity can be identified.

In cluster B, the common denominator of the three components is unit resilience. The connection between maintaining an organic framework, the burden of activity, and organization of unit belonging enables:

1. Direct and immediate reciprocal relations between the components.
2. Spreading damage or compensation. How is this possible? Damage to the ability to maintain an organic framework (B1) and/or damage to organizing unit belonging and identification (B3) may harm the ability to cope with the burden of activity leading to exhaustion (B2). In other words, damage to the sense of participation in the common effort, both mental harm and harm to morale, may create the sense of an enormous burden on the shoulders of the single soldier. However, if there is damage only to the ability to cope with the burden that leads to exhaustion, this may become more moderate if there is identification and cohesion with the organic framework and/or in the organization of a sense of belonging and identification with the unit. Put differently, the soldier will experience the sense of being burdened, but his identification and his feeling of belonging to the organic framework may prevent a feeling of oppressiveness and may lighten the burden significantly.

B. *Regarding the reciprocal relations between the clusters themselves: The connections between them are not immediate, and there are clear differences among each of them.* Any attempt to propose that they have direct relationships in the same way as, for example, the connections between the elements that compose each cluster, may miss the differentiation between clusters. Therefore, it may be assumed that spreading damage is possible between one cluster and another without possibility of compensation, as the passage between one element to another of the same cluster still leaves intact the common denominator of that cluster, which does not exist for another cluster. On the other hand, the differences between the various clusters may block the possibility of spread.

Another question that should be examined regarding the grading of resilience is: what happens in a situation of damage to all the elements of one complete cluster? Is it still possible to claim that damage to single elements of different clusters is more serious that damage to a complete cluster?

The idea that was raised about the system of compensation that exists to a greater extent among the elements of one cluster and does not necessarily exist in other clusters, emphasizes the severity of the possibility that damage will spread from one cluster to another. In other words, an extreme situation of complete damage to a cluster may create certain damage to another cluster. An expression of the severity of damage to all of the elements in a cluster is the complete absence of this cluster as a component of military resilience. How can spreading damage from one cluster to others be prevented? For example, complete collapse of military spirit would be a catastrophe in managing combat on any level, which deviates beyond a discussion of the elements of one cluster or another. And this is due to spreading damage.

In addition, from a methodological standpoint, it is important to focus on logical dilemmas in examining the reliability of a theory that is being raised here for the first time, with a possible scenario. A scenario such as the complete collapse of military spirit is extreme. Extreme scenarios are important and necessary in the realm of national security thinking, but these situations should not be able to disprove theories as the theories are usually put forward while taking into consideration any possible reality. The limitations of a theory may derive from an extreme realty that may improve the theory in dictating necessary conditions and adding to its validity, but an extreme situation cannot, in itself, disprove the theory.

A solution to the dilemma of the importance of the elements in relation to the clusters lies in the fact that military resilience as it is being discussed here refers to the human factor. It is logical to assume that the possibility exists for ranking the importance of military resilience in these terms, but, as the object of the research is human, the ability to isolate associations among the various elements of resilience will always be limited.

Low-intensity Conflict

As mentioned prevously, a low-intensity conflict can be the continuation of a full-scale war, a separate episode that develops and changes, and eventually in certain circumstances, may develop into full-scale warfare. A clarification of the complex nature of a low-intensity conflict with its asymmetric dimensions is at the center of diverse professional literature.[8] There is wide agreement regarding the shifting and elusive nature of the field of operations of the conflict, involving the civilian and the military in the structural, organizational, operational and psychological-moral dimensions of the irregular warfare of guerrillas or terrorist organizations, psychological warfare, sabotage, and open and secret terrorism.

This conflict poses new problems for political decision-makers and those who prepare military operations. They must now deal with both the adaptability and the mobility of an enemy who is inferior in warfare power. The combination of relatively simple armaments and technologies that have media impact, in the framework of a strategy directed by cultural principles (ethos) that are in opposition to Western values are at the basis of the "problem of asymmetry" of the conflict.

A low-intensity conflict has a political purpose, and its resolution is accomplished by a change in social awareness caused by long-term exhaustion. A low-intensity conflict is characterized by the following:

1. Political considerations in a low-intensity conflict are dominant while the military-operational consideration is secondary.
2. The main outcomes are those of consciousness while the physical outcomes are secondary.
3. Low-intensity conflicts are long and their moves are planned correspondingly.
4. The generator of the conflict resolution is the change in consciousness in society by means of manipulating opinion.
5. The management of the conflict is based on the ability to supervise the intensity of the friction.

A low-intensity conflict combines non-violent actions, such as civil protest such as strikes, demonstrations and disruption of order; along with violent acts, such as violent disruptions of order, guerrilla activity and terrorism. The forces participating are irregulars who do not belong to an organized military nor are they part of any regular army. The choice of a certain mode of action is made in accord with the measure of its contribution to achieving the goals of the conflict.

In his book *Small Wars: Their Principles and Practice,* Callwell,[9] describing the tactical and strategic fundamentals of low-intensity conflicts in the Imperial Age, refers to common failures in the conduct of this type of conflicts at the political and the military ranks, and analyses the asymmetrical characteristics of low-intensity conflicts at different levels.

Conflicts with different motivations have common characteristics: all take place in a basic situation of asymmetry in the power of the opposing sides. The asymmetry dictates the use of strategy and operational methods that provide responses to the basic situation. Thus, it is possible to amalgamate what has been termed up to now a popular uprising, guerrilla warfare, terrorism, civil disobedience, a revolutionary war, a war of subversion and others, as, fundamentally, they have a common nature and they are intended to achieve goals with strategic political significance.

As stated, what creates the nature of a low-intensity conflict is the great measure of asymmetry in the power of the opposing sides.

As a military-security concept asymmetry refers to the following:

- A situation of conflict between a side that enjoys generally greater abilities and strategic power, or technological abilities (that is, in defensive-military technology) and a side that is inferior in general power or in technological abilities.
- The above phenomenon is common and widespread in the realities of our times and in the defensive-military and political realities of the future. This indicates the fact that, in the existing and developing technical and political realities worldwide, and especially in the democratic- Western world, achievements can be made, outcomes and even victory may be achieved even with warfare methods that are not advanced if they are directed at efficient exploitation of the weaknesses of the opposing side while actually turning its advantages into disadvantages.

The idea of asymmetry is not new in the advanced national defense systems and armies in the present, and it is a consequence of the complex requirements in the field of national defense and military operations, which have developed in the post- cold war era. These requirements were a result of the reciprocal effects of the dominance of the democratic Western World and its up-to-date accepted concepts of modern technology, its outcomes, and its effects. Methodologically, there should be distinctions made between some aspects of asymmetry between the opposing sides:

1. Asymmetry in the basic national data-territory (geo-strategy), population, resources, power and economic ability, status and location in the international arena.
2. Asymmetry in national values, and among them: belief, ideology, perception of the importance of territory and national or tribal membership, the existence or the non-existence of binding traditions; basic national values in contrast to universal values, reciprocal relations between these and relative priorities; what is considered an essential national interest and what is not considered essential, the importance of national interest in contrast to personal interest and the status of the individual and his/her rights; the measure of commitment to all of these, the measure of decisiveness and readiness to make sacrifices for the general good, the measure of national steadfastness; what is considered loss and sacrifice, situations and operational methods that are tolerable or intolerable; sensitivity towards and the importance of national public opinion and the media, as well as towards influences, pressures and national cohesion and their force against the com-

petition, and against political, sectarian and ideological differences, and the existence of pressures of minorities including irredentist factors; the attitude towards state security in contrast to other interests, and the measure of readiness to invest resources in state security needs, and the likelihood that all of these will be put into practice over time and in the presence of a threat that is presently low; the approach to warfare and violence and attitude towards using them; and others.

3. Asymmetry in the types of perceptions and interpretations of phenomena, behaviors, expressions and types of national, social, and political reactions and drawing conclusions from all of these.
4. Asymmetry in the data of status, support, and international contacts, especially with dominant international bodies, particularly in the realm of national defense and its requirements.
5. Asymmetry in technology: In technological ability, in technologizing and computerization of national systems; in the existence of reciprocal relations and in provision of systems developed in the civilian sphere to the military; in the measure of investment and readiness to invest in technologizing national defense systems and the military; in the dependence of national and defense-military systems on technology and development.
6. Asymmetry in the amount of possible damage to the home front and its resilience, in the measure of dependence and reliance on the home front for building and operating military force; in the steadfastness of the home front and in the possible effect of its being harmed on the creation of military force in war, including calling up the Reserves and its effect on the willingness of soldiers and particularly reserve soldiers.
7. Asymmetry in the available military capabilities: Those capabilities that are recruited on a regular basis and those that are manifested in the Reserves, and afterwards, consideration of the length of service and the potential for achieving readiness for action in terms of the numbers of soldiers, their equipment and the various military systems.
8. Asymmetry in the human, personal and group military abilities and qualities: Courage, discipline, willingness, readiness to sacrifice, battle spirit (their existence, conditions to create them, their strength, their durability); methods and capabilities of command and control, and in this context, the existence of initiative and relying upon it; level-headedness, cohesion, functioning under pressure, adaptability, the ability to absorb and withstand; the effect of the general state of the nation (difficulties, injuries

> and damage, public opinion); the measure of obligation to the military and its functioning; self-confidence, faith in the army command; quality of military command, and especially the high and medium level command-steadfastness under pressure, professional skill, obligation, knowledge and intellectual-professional flexibility; the professional level and professional skill in the defense system and the army in general.[10]

The description of these clusters challenges the popular notion of the "asymmetrical conflict" that has been usually discussed in military terms, and refers to states and national conventional military organizations as more dominant than a non-state actor. These challenges propose that between opponents such as state actors and non-state actors the notion of asymmetry is dynamic. In some clusters a non-state actor, such as a terrorist group or a guerrilla organization, owns a dominance on the other side, and the asymmetry is in the benefit of the non-state actor. In other clusters—the benefits belong to state actors. An additional challenge is the dynamic aspect of asymmetry due to fluctuations in these clusters for each side.

The U.S. Army's training manual FC 100–20 defined low-intensity conflict as "a limited politico-military struggle to attain political, military, social, economic or psychological objectives. It is often of lengthy duration and extends from diplomatic, economic and psychological pressure to terrorism and insurgency. LIC is generally confined to a specific geographical area and is often characterized by limitations of armaments, tactics and level of force. LIC involves the actual or contemplated use of military means up to just below the threshold of battle between regular armed forces."[11]

This definition is relevant as it includes most of the characteristics of a low-intensity conflict and the dynamics of its process while, at the same time, the principle of asymmetry of military means remains.

The Independent Variables: Military Model, Warfare Strategy, Political-Military Relations: The Relationship between the Military Model and Its Resilience in Low-Intensity Conflict

So far, the literature discussed the military model while examining the trends occurred in the nature and structure of the world's armies in light of technological changes, cultural, strategic and others. Discussion on various models were held and developed around the world, trying to understand the factors that led regimes and various companies to prefer one or another model.[12] In a historical perspective, the military model is not a static concept but a dynamic, and its design processes are always changing according to changes in its environment.

From a Modern Army to a Post-Modern Army

Charles Moskos,[13] argues that Western armies have recently shifted from the modern military model to the post-modern military model. Moskos distinguishes between three different periods characterized by different trends in different models of the world's armies: the modern military model took place between 1900–1945 (despite the fact that its roots and its origins could be identified in the late 18th century and especially the army of Napoleon); the latest modern military model took place during the Cold War, from 1945 to 1990; the post-modern military model is product of our time, beginning from 1990 (Moskos ibid.). The following table summarizes the aspects of the three models:

Table 1.1 Armed Forces in the Post-Cold War Era (Moskos 2000 p.8)

Armed Forces Variables	*Modern 1900–1945*	*Late Modern 1945–1990*	*Post Modern (since 1990)*
Perceived threat	Enemy Invasion	Nuclear war	Sub-national and non-military
Major mission definition	Defense of homeland	Support of alliance	New missions, e.g. peace-keeping, humanitarian
Public attitudes towards military	Supportive	Ambivalent	Indifferent
Media relations	Incorporated	Manipulated	Courted
Conscientious objection service	Limited or prohibited	Routinely permitted	Subsumed under civilian

Moskos sees the post-modern army as characterized by loose connection with the nation state; relies on volunteer force; diverse missions beyond the framework of the war, based on advanced technology and expertise; his ethos composition more varied (both in terms of value, both ethnically and in terms of integrating women); a weak distinction from civil society; motivation of the soldiers is no longer related to, or at least not exclusively, patriotism, but to a self-realization with a meaningful personal experience.[14]

This study chooses Moskos's analysis on models of armies as it hits most of the military linkages at its internal and external aspects. In addition, an exhaustive range of aspects is relevant—i.e., the structural aspect (e.g., force structure), functional (such as dominant military professional, etc.)

cognitive—mental—cultural aspects (threat perception, public opinion, conscientious objection).

The attempt to understand the influence of the military model on military resilience in low-intensity conflict arises for the first time because of contradictions between the components of the post-modern military model and the conduct of a low-intensity conflict.

The conduct of a low-intensity conflict in the postmodern era, which is seemingly consistent with the perception of the threat in this era (subnational and non-military), is inconsistent with the force structure (based on advanced technology and specialization). The analysis of asymmetrical properties of the war against terror / guerrilla emphasizes the limitations of a force structure based on advanced military technological sophistication. Fighting a low-intensity conflict is characterized with tactical maneuver, contact fluidity between the parties, less strategic machinery and emphasizing the human factor.

This does not mean that raising this contradiction implies that military technological sophistication is not necessary in introducing advanced information and applications for the military service in such conflict management. However, this is not the kind of the technological sophistication threshold relevant to conventional and non-conventional battle scenes. Moreover, the force structure characterized by high technological revolution requires extensive streamlining and reducing manpower. The historical experience in managing low-intensity conflict has proved that it cannot afford reduction array fighters. When confrontation was conducted away from home, armies needed greater array fighters, which allowed their flexibility against the high mobility characteristic of the terror and guerilla warfare, in addition to ensuring the continued regular military presence at home. Moreover, despite the threat perception of the Post-Modern model, managing a low-intensity conflict can be at the same time while conducting a management of regular conventional battlefield with other opponent states. Besides the fact that a low-intensity conflict might be a development of insurgency that stands for itself, a low-intensity conflict might precede a full-scale warfare, and it also might follow a full scale war. Therefore, the threat perception of the post-modern army is not necessarily consistent with small professional armies.

Erosion factors and the extent of time that characterize the threat perception of the post-modern model, that is, the prolonged warfare required in the conflict against terrorism/guerilla, do not get their support from the post-modern characteristics of indifference in public attitudes towards the military, courted media relations, and a conscientious objection service that is subsumed under civilian. Media relations are no longer communication mobilized for national goals and or given to maneuver. This is a demanding media with needs to invest resources in military courtship. The conscientious objection may have implications on the array fighters that conduct a war, which by its nature poses a moral challenge not small.

If so, the post-modern military model is not "friendly" to military resilience in managing a low-intensity conflict. The characteristics of this type of conflict pose critical challenges on social resources, national resilience and cohesion. In the psychological warfare that stands in the center of this type of conflict, the post-modern military model does not guarantee social environment support. This model's loose ties with the nation-state, dependence on volunteer force, weak linkage with civic society etc. make it difficult to compliance with the worn long time dimension of low-intensity conflict.

This problematic aspect of the model is the starting point for the assumptions attributed to study the relationship between the military model and low-intensity conflict management.

This setting of the post-military model works for the states discussed in this book even if in practice it is a proto-type model that does not exist during their conflict management. However, in all the states the military model is in a transition process from one type to another, when the general trend is towards the post-modern army. Precisely because of this, it is allowed to stand on the validity and limitations of the model hypothesis concerned with military resilience in a low-intensity conflict.

The Relationship between Military Strategy Warfare and Military Resilience in Low-intensity Conflict

In analyzing the warfare characteristics of low-intensity conflicts Harkabi[15] notes that conventional warfare strategy, which is the central classical purpose of the army, is not relevant to the management of anti-terrorism and guerilla warfare. This is a combat strategy with blurred lines between combatants and non combatants, when next to instrumental factor, i.e., the instrumental character device there is a dominance to the expressive factor, i.e., the social-emotional character linked to self-expression-related emotional and symbolic ceremonies.[16] A low-intensity conflict is prominent with an asymmetry between the warring parties that marks the circumstances of this warfare and its tactical methods. That who fights this war usually does not take known parallel warfare methods, namely guerrilla against guerrilla and terror against terror, but runs conventional army units that adapt a fighting strategy, which is incompatible with the formal designation of armies in conducting regular warfare. Whole units are required to perform the type of activity ongoing role different from designated wartime defined official purpose.

The characteristics of low-intensity conflict produce more war principles for the military. These are gaps between the mission of various military units and the demands that fighting against terror and guerilla organizations set for these units. Based on this logic that sees assigned missions a type of practice that produces meaning in its wake, strengthening the resilience of the array warrior, the hypothesis about the relationship between warfare strategy and military resilience is proposed. The type of strategy applied in this fight affect military resilience at the individual level and organizational level. This position is important for understanding early review of possible ways in which warfare can enhance resilience[17]:

> *At the individual level*, most of the regular soldiers recruited at the age where self-esteem is not crystallized enough, their inner world can still be stormy, full of drives and impulses that typically do not receive an external expression. This internal dynamic is a platform to motivate soldiers to fight the battle even if this is performed in an unconscious level. In terms of fighters, a situation of fighting is a legitimate channel to direct energy to allow satisfaction of those instincts. For them, the entrance to the field fighting, together with the fact that their unit was chosen to do so from a variety of units, is an important component in formulating their self-esteem. Desire of warriors to take part inaction testifies to the presence operative action indicates coping resources and mental powers as well as openness to experiences that apparently can prevent the formation of "byproducts" negative due to the war.

Motivation to out activities is conditioned on several factors:

a. Activity must be perceived as unique, with quality, necessary and perceived as such that a chance of fire contact should be higher.
b. The fighters need to feel that the activity is estimated both by elements of the military and outside the military.
c. Activity should be seen as dynamic, and as such that contains elements of constant motion.
d. Activity should be seen as successful and contributes to the overall effort to thwart terrorism.

In light of this, we see that as long as the fighting takes place according to these parameters, there exists among the fighters a sense of meaning, mission and impact on events that in turn affects their morale and motivation.

On the other hand, when the fighters are required to defensive activity characterized by a markedly passive security such as checkpoints and ensuring communities, their motivation level goes down as well as their ability to cope with fatigue, erosion and low morale. Support for this insight is in Weiner's article[18] where a comparison was made between Israeli combatants on the northern border area and those in the West Bank during Operation *High Tide Low Tide*. The findings indicate that commanders and fighters feel better in northern blue line than those in the West Bank. Weiner suggests that unrealized threats in blue line produces similar bureaucratic discourse similar to the one in daily military routine in order to address the threat when this comes to practice. On the other hand, in the West Bank the realization of the threat breaks the bureaucratic life, and organizes the activities in what he calls "combative discourse," which produces an atmosphere of mission and meaning for soldiers in the field.

This assumption is also supported by the difficulties experienced by combat support troops. On the one hand, they are needed for intensive work on which the fighting force is based. On the other hand, the fact that they are located outside the cycle of violence does not allow them to share their assessments praise harvesting as the other fighters. On this background, combat support troops experience a sense of marginality: theirs and their performances vis-à-vis the operational activity.

At the organizational level the prolonged fighting is an opportunity for the military to learn on "wet" while being exposed to a variety of events that require new responses that were not included in traditional combat doctrine and learning from them. Activity itself is the best workout. The exchange of forces also allows the passage of knowledge from one unit to another formally and informally acquired. In addition, the very existence of the turnover makes erosion less likely to occur and fighting routine lessens. Furthermore, both officers and commanders, are given the opportunity to lead troops, experience command, make decisions and deal with the complex dilemmas otherwise would not have to deal with. This is particularly true for junior officers that due to dispersion forces over a large area are given a chance to express their capabilities command in a way that prevented them in other combat situations or in routine times.[19]

The ideas in favor for fighting as a contributor to resilience are restricted by several factors, including the question of the purpose of fighting, for example, how far are the required combat operations consistent with the formal designation of the array warrior. The following definition of *warfare strategy in low-intensity conflict* is based on this emphasis: "Conduct of ongoing operative activity different from the role designated for unit forces in wartime and has been their defined official mission."

The Relationship between Political-Military Relations and Military Resilience in Low-intensity Conflict

Political-Military relations constitute a complex dilemma in the literature that sees a dialectical relationship between policy and strategy usually normative in character. There is an agreement on the dominance of policy on strategy, when war is considered as an instrument of policy, and strategy is a test of the instrumental rationality of the policy. This agreement is accompanied by limitations, such as when the policy sets tasks to the military rank it should take into account considerations of what can be achieved by the army, and the price for this achievement. In addition, although strategy serves means to policy, it also constitutes a criterion for goals implementation, and points on difficulties on their the way to realization.[20]

The theoretical development in this dilemma of political-military relations describes the historical development of relations between the two ranks. It has always been a structured conflict due to reasons such as:

a. The military rank seeks for totality in fighting; political leadership seeks to relativity and restraint.
b. The military seeks a clear definition of the objectives of the war; political leadership seeks for ambiguity.
c. The military seeks for freedom of action in fighting; the military seeks for continuous control.
d. The military sees war as policy failure; political leadership sees it as policy continuance.

Apparently, it seems that since in the conduct of a low-intensity conflict the military level is more characterized by transition of strategy to tactics, it seems as if the demarcation line between political and military leadership is more clear. Especially since this military conflict and its management is usually not part of a regulated political program, as is usually required for the management of regular war. However, in practice against the background of this luminosity, in which the two ranks are closely linked in an extensive cooperation (low-intensity conflict management efforts are coordinated between the two levels in most areas: political, security—military, economic, diplomacy, international law, Media, and so forth) the boundaries between them become blurred towards militarization of politics and politicization of the military.

One cannot ignore the essence of civilian control on the military due to its relevance to managing low-intensity conflict. The key issue in this control is the civilian instructive aspect of political leadership to the army, that stems from seeing the civilian control as a function of accountability. Civilian control can be purposeful only when citizens can demand from army officers to bear responsibility for their loyalty to the regime, what they say, what they do and they did. Control is made possible through effective mechanisms of accountability that enable civil authority to require military officers to bear responsibility, to examine them carefully, investigate them, and order them or restrict their behavior in cases of exceeding the authorized standards. When these mechanisms are strong and effective, civilian control is strong and effective. Civilian control is the means to mark regime rules, which defines the rules for all players, outlining mechanisms of accountability that should allow control over how authority is being practiced and how to play the game. Civilian control is the essence of the division of responsibility between civilian and military, while civilian authorities should monitor the goals and policies concerning national security resource allocation and use of force, and the army has some degree of authority about military doctrine, discipline, operational plans, internal organization and promotion of officers and operation of unit tactical operations.

The relations between the two ranks in the post-modern era place additional structural and functional challenges that could increase all the tensions discussed so far. These stem from the profile of top military officers in the post-modern army.

New tasks and changing the military model also led to a change in the person of the commander, both in the public eye and from the perspective of the military organization. The "manager" or the "technician" officer" do not give way to the "diplomatic" officer or to "rescue worker soldier," but rather complement them. Preparing an army for a humanitarian mission or peacekeeping mission in distant countries requires many changes, including goal setting, the threats and risks, defining the enemy, mental and operational preparation of the soldier, consideration of various foreign powers and local culture, etc. These bring a change in the form of the commander. A commander is no longer just "combat leader" that look at the data to the battlefield to achieve victory, but also "Commander-Diplomat,""Commander-communication," "Commander-Scholar," and "Commander-Statesman" entrusted with political contacts management, bureaucracy, contact with foreign populations, communication management and marketing national mission among the public and coordination between civilian organizations.

Two more images characteristic of modern army personnel that own close relationship with the scientific trend of the army and its technological and administrative complexity:

a. The "soldier-scholar"
b. The "soldier statesman"—also seen as "soldier- politician," mainly because senior officers involved in political debates concerning the security policy.

Against the background of this profile, which could challenge the political leadership skills, because those already included in the profile of the post-modern officer, and the unique political and military characteristics of low-intensity conflict, there is an increase of sensitivity in the relationship between political and military leadership. At the same time a lack of frequent and powerful non-consensus, which could damage military resilience within the conflict perception and military conduct.

On this background the definition of the relations between the two ranks in a low-intensity conflict concerns the nature of communication, how they work vis-à-vis against each other and how the interface between them is managed and regulated on the background of the structured conflict between them.

Low-intensity conflicts possess crucial political, social, economic, regional and global importance. Comparative research between countries that have conducted and/or managing such a confrontation hold a theoretical contribution to the problem of military resilience. This understanding may also promote research in practical terms.

Chapter 2

A Trilogy of Paradoxes and Their Mode of Analysis

The research issues this book discusses can be described as paradoxes deriving from the relationship between military resilience in low-intensity conflict and variables, which, for the first time, are considered to have an effect on military resilience. These variables are *military model, warfare strategy, and relations between the political-military ranks* in the conduct of the conflict. Three paradoxical ideas provide a basis for understanding the hypotheses to follow:

1. *A post-modern army*, based, in essence, on missions characterized with low-intensity warfare, on the one hand, and high technological professionalism on the other, has been discovered to endanger military resilience. High technology–oriented armament and training (for full-scale warfare (conventional and nuclear) present a non-relevance to conduct low-intensity conflicts.
2. *Low-intensity warfare strategy* endangers military resilience
3. *Relationships between political-military ranks* bear greatest tension in managing a low-intensity conflict, in spite of the fact that, in this type of conflict, military operations are characterized by the transition from strategy to tactics, and management of the conflict is not usually part of an organized political plan. The transition from strategy to tactics is required due to the constraints that this type of conflict places on both political and military ranks. In other words, the means impose limitations upon the objectives, despite the subordination of the inverted hierarchy.

In this type of conflict, there is no importance to the dialectical link between the political level and the military-strategic one concerning the various considerations made usually within the classical mission of armies in full-scale warfare. This dialectical link is being reduced; the linkage between political-military is being detached in its transition in the fighting at a tactical level. This is a level where specific incidental battles are being repeated, often without mutual linkage to a big battle. This transition is dictated by two contradictory factors:

- The rationale of irregular warfare is stated attrition, which seeks to achieve its goal by the accumulation of attack operations. The organization that controls irregular warfare faces a problem of how to switch from a cluster of accumulated attacks to an orderly sequential multi-stage plan leading to achieve the purpose. Lack of such sequentially in fighting low-intensity conflict diminishes strategic aspects in favor to tactics.
- The rationale of classical-regular warfare sees strategy as the overall, and tactics a part within that. Thus, strategy is to lay tasks on tactics, not vice versa. Hence, the transition of strategy to tactics is forced by constraints that low-intensity conflict poses to the political and military ranks. In other words, the means lay limitations on aims and objectives in spite of the opposite hierarchical subordination.[1]

I. Structural, functional, and social-mental factors of the military model affect the resilience of the army.

> *The more the military model is post-modern, the greater chance there will be of damage to resilience in a prolonged low-intensity conflict.*

This hypothesis is based on the characteristics of the post-modern model. An army that, on the one hand, holds a threat perception based on the salient threat of internal crisis or a non-military threat and on the other hand, whose structural dimension is based on the concept of an advanced technological professional army, there is a gap created between the perception of threat and comprehensive training as a response to this threat. In this gap, a sense of damage to the mission of the warfare system in all its aspects is created. The absence of a correlation between sub-national conflicts and threat perceptions emerges also from other structural characteristics of the military model such as high military technology especially dedicated for future battle scenes, and fosters irrelevance to this military technology dominance. This gap requires a process of especially dynamic adaptation of army structure to threat perception. Adaptation as an enduring lasting process is usually transformed into erosion. Against the backdrop of increasing tension in this process, there are greater chances of damage being done to the building of force, to unit resilience, to the military framework, to significance in the realm of military spirit and to operational fitness.

Even in cases where technology and specialization are directed towards dealing with sub-national conflicts, such as terrorism, this contradiction is still valid for two major reasons:

1. The diversion of technology and expertise in the post-modern army to deal with such conflicts should not be taken for granted. Factually, specific technological developments to deal with war on terrorism have an ad-hoc character (which for the most part is prevented by business considerations) and ranges from intelligence technologies such as indentifying suspicious military objects to tactical-offensive capabilities such as to neutralize explosive or explosive-laden individuals. Non-lethal weapons that offer a technological solution to dealing with terrorism are limited to specific situations, when the site of the terror attack is defined, territorially accessible and the military-civilian are not involved in confusion. The scene of the battle against terrorism is usually not characterized by these qualities at its onset stage. The salient characteristic of technology of the post-modern army is the ability to fire precisely from afar against the opposing side with the capability to absorb return fire. Firing precisely does not enable coping with a low-intensity battle arena in populated areas; the ability to distinguish between this technological power and its influence on consciousness has not yet been formulated due to effect that it creates in surprise, in magnitude and in the extent of fire power. Unique technological developments for dealing with terrorism are still limited in their ability to be utilized in civilian populated areas.
2. The non-congruency between threat perception and advanced professional technology does not invalidate the importance of the technological role in managing a low-intensity conflict. However, physical access to a territorialized terrorist organisation is not always possible due to its evasive character. This book is very much aware of the transformation of a low intensity conflict, and sees the gaining of physical access to terrorist entities as a developmental stage and transformation in managing a low-intensity conflict and in counter-insurgency. In other words, when such entities become accessible and physically delineated, they lose their relative advantage of being evasive. Thus, their identifiable nature makes them more accessible to military technology. This is the maturation stage of the conflict in which there is perhaps a place for the advanced technology of the post-modern army. This book intends to elaborate upon the development of this dimension in the states that will be discussed.

In the maturation stage of a low-intensity conflict, there would seem to be a solution to the non-congruency raised here. It is not in the development of advanced weaponry, but rather the maturation of a conflict that leads to the use of completely advanced conventional technology.

As for the other aspects of the post-modern military model: lack of supportive social-military relations, media-military relations along with conscientious objection arouse mental dissonance in combatants, and increase the probability to damage military resilience, especially in Unit Resilience, Significance in the Realm of Military Spirit, Significance in the Realm of Operational Fitness.

Hence, a post-modern army will have difficulty to maintain its resilience in low-intensity conflict. The characteristics of the post-modern military are not conducive to conduct such conflict:

- Due to the loss of the national army ethos grounded in its characteristics
- Dominant civil affinity that reduces the gap between army and society and violates the borders of the army as an institution
- Critical humanitarian liberalism not necessarily consistent nor with the challenging missions armies face in such conflict neither with the endurance needed in such conflict

II. Warfare strategy in a low-intensity conflict affects the resilience of the army.

> *The farther the warfare strategy is from the army's defined mission and from formal military warfare training, the greater the chance that military resilience will be harmed.The more the warfare is managed as small-scale terror-guerilla warfare, the greater the chance of harm to military resilience.*

The hypotheses rest on the problem of the asymmetry in the conduct of a low-intensity conflict. Even more than the damage to the mission when managing the fighting, the expressive dimension (military presence in a civilian area, emphasis on the human factor in fighting rather than the mechanical factor, fighting at close quarters, among others) creates a blurring in the identity of the soldier-citizen. Its expression is well-known in that it creates mental dissonance that causes damage to the areas of corps spirit, unit resilience etc. This means a disruption in the stability of the effective implementation of the military alignment. Preparing the military alignment for its mission of limited combats does not necessarily change the character of the battle scene and its expressive-dynamic load. That is, even when specified units are prepared for small-scale battle, this should not be seen as a design for the subjective perception of the fighter, taken for granted as stable over time, against the backdrop of arena conditions that are so demanding. The operational-combative dimension in this preparation may ensure stability on the level of fighter performance, but the dominant expressive character of a low intensity battle scene against a terrorist group within a civilian territory may create cracks in the mental dimension due to the instability and the changes in the subjective perception of the dynamic environment. These cracks may disrupt the effective operational stability of the fighter.

III. Relations between the political-military ranks in a low-intensity conflict affect military resilience.

> *The absence of a consensus between the political and military ranks regarding the objectives of the conflict and/or the means to manage the conflict increases the chance of harm to military resilience.*

Clear dominance is attributed to this hypothesis.

For soldiers and combattants the relations between the political-military echelons embodies the granting of legal authority, by which a consensus between the two echelons actually supplies legal and moral legitimacy to the goals of fighting in the conflict and how these goals are to be achieved. This is the source of decisiveness of both the military and civilian leadership, which, at the highest levels of command, supplies a basis for the attempt at explanation and persuasion that are taken in the various units. Regarding the level of the individual soldier: leadership consensus supports the internal persuasion of each soldier regarding the justice of the conflict's *objectives and the means to conduct it*. However, the influence of the consensus particularly has a moderating effect in managing the mental dissonance that may occur at any level of the military warfare alignment, in the absence of a consensus on the goals and the means used in the conflict.

Regarding the mental dissonance on the level of the individual soldier and/or group and the dissonance on the level of military command, it is considered that the possibility of the latter dissonance emerges against a background of tension between the military and the political echelons. Although this has ramifications on the personal level of the soldier as well, *great damage in military resilience occurs when the dissonance is at the level of the senior command.*

It is maintained that failure to solve a lack of consensus between the political and the military echelons and its deepening as time passes, can even lead to a split in the military organization itself, leading to immediate and maximally great damage to its resilience.

The discussion of these hypotheses will be performed in an analysis of professional literature on the military of several states involved in low intensity conflicts:

- Historical sources
- Documentation of battles and testimony on various levels
- Newspapers and Internet
- Military documents that have been released for publication
- Data on the military model in the countries being investigated at the time of the conflicts are based on relevant professional literature:

The structural and functional dimension of the model in each state (the structure of the army, the missions of the army) will be taken from literature that examines changes in modern armies and problems of army reorganization.

The perceptual, cultural-social dimension (the perception of threat, relations between the media and the army, relations between society and the army, and conscientious objection) will be derived from professional literature dealing with how these were reflected in each conflict.

The same sources will be used to investigate the political-military echelons in each state during different periods of the conflict.

Conduct of the battles and military operations that are documented in the military history literature of each conflict will supply data on the warfare strategy that was adopted and what its orientation was.

Regarding military resilience: in addition to the factual documentation that exists to one degree or another in every country, an additional interesting and especially relevant aspect of the necessary data will be taken into account. Testimony as witnessed by the fighters in the field will provide information about battle conduct. This testimony is especially authentic as it documents *the process of mental changes of the fighters themselves, changes that they experience before their military involvement in the conflict as they reach the conflict area, during the conflict and years after its end.* This data is available for the French-Algerian conflict, and the British-Irish conflict. Data can also be gathered about the Russian-Chechen conflict and the Israeli-Palestinian conflict, but, as these conflicts are still going on, sources of data will be primarily based on current information such as newspapers, Internet and the limited research literature.

A post-modern army, warfare strategy and relations between the political-military echelons will be analyzed through the following aspects:

- *A post-modern army*: perception of a non-military threat (a sub-national and a non-military threat), small, professional military with a fundamentally humanitarian goal. The public attitude towards the army is indifference; the army courts the media and it anchors conscientious objection in civilian service.[2]
- *Warfare strategy in low-intensity conflict*: geographical conflict characterized by blurred borders, fluidity of close contact; emphasis on the human factor in the realm of people-machines, involvement of civilians, asymmetry in the coping methods.[3]
- *Relations between the political-military echelons in a low-intensity conflict*: consensus/absence of consensus in conflict goals, by justification of means of conducting the conflict (political and/or military) and by the type of warfare methods employed.
- The connection between the military model with all of its components and the military resilience will be tested as follows: The perception of threat, the structure of forces, the major mission definition, the dominant military profession, the relations between the army and society, the relations between the army and the media[4] and conscientious objection[5] will be identified in France, Britain, Russia, and Israel. The purpose is thus to classify the army in each of the countries according to the Moskos Table of Models[6] presented above and to assess which military model existed in each country while it was conducting a low-intensity conflict.

To the extent that the military model in each country will be characterized by the post-modern model, we will expect greater damage to military resilience in the conduct of the conflict.

The warfare strategy of the countries involved in the conflict will be traced and development over time will be analyzed. To the extent that warfare in the conflict is characterized as small-scale and stable over time, a greater damage to military resilience is to be expected. Put differently, to the extent that warfare strategy is characterized by the following aspects, a greater damage to military resilience will be expected:

- *The geographical conflict* is characterized by blurring of borders and fluidity of contact, even in close warfare. In contrast to conventional warfare, in which the conflict takes place at the front lines, in low-intensity conflict the front line is indistinct, as front and rear are mixed, and the concept of the front line is not applicable. While in conventional warfare, there is emphasis on maintaining contact, terrorists and/or guerillas aim for short-term and unstable contact, and they prefer to cut off contact and keep their advantage in the initiation of future hostilities. As they are the inferior side in warfare capability, they avoid clashing head-on for an extended period with an army, maximizing the damage they cause and leaving the decision and the initiative about when to expose themselves in their own hands. This is the rationale of "you can't win against an enemy who is everywhere and at the same time is nowhere . . ."[7]
- *Man-machine* is characterized by emphasis on the human factor. In contrast to conventional warfare in which there is a combination of man and machine in warfare, in a low-intensity conflict, the human factor is salient.
- *Professionalism as opposed to non-professionalism*—Involvement of civilians.

In conventional warfare a trained professional army is activated. In contrast, terrorists conduct a "war of the people" in which there is an intimate link between the fighters and civilians to a greater extent than in any other form of warfare.

Coping Methods—Asymmetry between the sides in the means of warfare management, parallel to tactical maneuvering with little fire, sometimes without mobility.

While in the conventional war model, there is symmetry between the two sides, one conventional army against another conventional army, the model of a low-intensity conflict is mixed: terror tactics against a force whose power is on a conventional level, using sophisticated means of warfare management that enable high fire power with strategic maneuverability and wide tactical mobility.[8]

The book focuses on a one-way examination of warfare strategy on military resilience, that is, the proof of damage to military resilience as a result of warfare strategy characterized by small-scale warfare components. However, in order to provide validity to this causality an analysis will refer to the transition from a small-scale wafare into a regular military warfare as one that leads to a recovery in military resilience.

As for the relations between the political-military ranks: a conflict managed only by lack of consensus in the means of managing it (either political and/or military) and in the type of methods to manage it, while there is consensus about the goals of the conflict, a low degree of damage to military resilience is expected. In the absence of consensus about the goals of the conflict, we will see a high degree of damage.

The communications aspect of management of tension between the echelons has a decisive impact on military resilience.[9] This issue will not be discussed, although this aspect cannot be ignored. However, the distinction made regarding consensus or lack of consensus about the actual objectives of the conflict can stand alone regarding its crucial relevance, and may therefore have severe implications on army resilience even without the mediation of mass communications.

Here too, the one-way principle will apply in proving lack of consensus between the political and military echelons regarding damage to military resilience. However, there will be reference to recovery in military resilience when there is consensus between the echelons, but not in great detail.

The general concept of causality aspect is elaborated in the following question:

How does each variable affect military resilience and what specific effects are derived from each one? In other words, if data is found regarding damage to military resilience, how do we know if this damage is a result of the military model? Warfare strategy? Relations between the military and political echelons?

This study posits that, in order to derive causation for damage to resilience, it is possible to apply understanding of resilience in a biological-organic system to an organizational-command system like that of an army. There are a number of similarities between the two systems for understanding resilience and the elements that affect it. Resilience in the two systems is dependent upon:

- Structural-physical-genetic conditions and environmental conditions
- Conditions of emotional-mental data
- The dynamic of the function of the individual/organization, the character of its activity and its ability to cope with external input using given resources
- Reciprocal relations among the above

Logical conclusions of this comparison for control of the variables are as follows:

Even if it is possible to isolate the effect of each independent variable, the isolation is meant to differentiate between the variables without ascribing explanatory exclusivity to any variable. This is due to the fact that the effect on resilience may occur primarily from the combination of factors that have a different effect of intensivity in relation to time.

In comparison to elements that affect the resilience of a biological-organic unit, the *military model* actually causes effect as a combination of *anatomical-structural-functional and mental properties* due to the variety of elements, such as the structure of the forces, the type of mission, the social-military relations, and others.

Warfare strategy has an effect as the *function* in which the organization is involved. Its nature is dynamic in coping with the external inputs using the given resources. On this level, it can be said that these resources are actually derived from the *military model*, for example, how big the army is, whether society supports army activity, etc.

Relations between the political and military echelons affect all levels:

- A wide structural effect that ties the army to the environment that is common to the dominant system that also has a monopoly on power, that is, the political system. This system also monitors the military function.
- The relations of the political and military echelons determine the "raison d'être" for the activities of the military organization and supply the legal-moral anchor for the codes of military function.

Against the backdrop of the reciprocal relations between variables in their effect on military resilience, data processing performed in this book tries in a specific and limited way to isolate the effects of each variable, as necessitated from the differentiation between the variables. However, usually it will be possible to be satisfied with a causal relationship, even if an effect on resilience in the presence of one isolated independent variable or more can be proven.

In the data processing, *military resilience*, will be presented in five different clusters: *building force; unit resilience; military framework; significance in the realm of military spirit; and significance in the realm of operational fitness*. Considering the reciprocal relations between the Military Model, Warfare Strategy and Politico-Military relations in their effects on *military resilience*, the effect of each variable is possible on each of the clusters of the *military resilience*.

The comparative research between the states examined here will be carried out on two levels:

- A comparison between states on the level of their *military resilience* and their military model, warfare strategy and relations between political-military ranks.
- A comparison between different periods of the conflict in each state.

The comparison is not sufficient to prove causality between variables. The role of the comparison is to grant validity, strength and reliability to the hypothesized causality between variables, that is, granting the causality its explanatory strength. This causality will be proven using inductive reasoning, which will make it possible to accumulate events that have damaged resilience and have been attributed to the presence of the three variables in all of the states, and in each one separately. From a methodological standpoint, the causal limitations derived from comparison do not contradict the possibility of inductive reasoning. Causality would be possible with research on one isolated state using inductive reasoning with all of the evidence found as supporting and strengthening the theoretical hypotheses. However, the comparison between states supplies a quantitative measure accumulated for what can be learned from several particular states. Thus, the validity of inductive reasoning would be inestimably greater against a background of comparison between different states. And if we are dealing with the formulation of a logical universal law that requires induction, the comparison may, on the one hand, provides the limits of this law and, on the other hand, defines its areas of validity.

Chapter 3

France-Algeria 1952–1964

The cases discussed in this book present a challenge for understanding the development of a low-intensity conflicts as a phenomenon. This is not only because of what can be learned from the chronology of the first case (France-Algeria, 1954) continuing up to the fourth case (Israel-The Palestinian Authority, 2004), but also because there is no low-intensity conflict that is similar to any another. The essence of understanding the challenge can be attributed to the difference in strategic cultures on the perceptual and the instrumental levels of each of the sides during the different periods, in both exogenous and endogenous aspects, as well. The validity of the comparison between the states is based on the assumption of this difference among the cases, and points out similar trends despite the differences.

France Military Model: The military model from 1954–1962 was the model of a late modern army.[1]

Threat Perception: Enemy invasion—atomic war

Threat perception in France had always been affected by the real and immediate threat to French territory and to the assets of the French empire throughout the world. However, after the Second World War, the Cold War, and the process of de-colonization, threat perception changed, with the fear of Communist army invasions and atomic war becoming more dominant in security thinking agenda.[2]

The conflict in Algeria should be viewed in relation to this threat perception: France viewed the African continent, and specifically Algeria, as having high-level geo-strategic value. In other words, it provided strategic depth as an operational arena that would enable retreat, redeployment and reattack. The Second World War proved the strategic status of Africa for France (It was said that without Africa, France would not have been among the victors). In addition, Africa provided France with raw material of strategic value both for threat perception and for its plan of defense.

The defense of Africa in general was linked to that of France not only as an integral part of the strategy and security of France, but as having a preferred status in its defense agenda. In light of the Soviet-atomic threat and the balance of terror, the future of the African interest was critical to the security of France.[3]

Force Structure: Mass Army–Universal Conscription

France was the first to use the national army model or a mass popular army. For many years, the military establishment in France defined the army as a widely-based military force relying on universal conscription at the highest rates, from among all age levels of regular army service and army reserves.

In the French-Algerian dispute there was a large force structure—about a half million people (about half of them, conscripted to compulsory army service), and with the reserves, a million people. In addition to the army, there was an army-like police force for internal security missions (in France and abroad), the gendarmerie. France also had its Foreign Legion that had been established in 1830. Its mission was to assist France in governing its colonies in Africa, and afterwards in Asia and in America, without endangering the lives of the French. The attraction for joining the Foreign Legion was based on Legionnaires seeking adventure, interest, professionalism, a fair wage, and especially, the fact that each conscript received a new identity, prestigious status and in the end, a French identification.[4]

The force structure is derived from the perception of the importance of defense of the colonies. Since the reworking of French national security perceptions in January 1959, every formal, legal, budgetary, political and strategic document emphasized the centrality and the importance of involvement in Africa. To this end, the military structure was directed to fulfill obligations both to the center: an army with atomic power, and to the periphery: a defense force for the territories.[5]

Algeria became the place where the red line would be drawn for French concessions in its colonial wars: no more retreat, no more defeat. As a national interest, at least at the beginning of the conflict, this granted the military model its status as an "an army of the people, convinced that the continued existence of the French empire was an imperative condition for safeguarding the status of France as a power." Calls for unity of the army and the nation, and for civilian obligation were clearly heard at the beginning of the conflict with Algeria.

Central Mission: Homeland Defense—Support of Allies

The central military missions were support for allies (NATO) and ensuring homeland defense against a Soviet atomic threat. The use of conventional forces was reserved for situations that demanded limited time involvement and were outside of the European arena, especially in black African states.[6]

Three aspects may be noted here:

1. The military mission that was imposed on the army in managing the conflict in Algeria also included a civilian-psychological dimension—"calming Algeria," that is, defense of the civilian population against the rebels and rehabilitating the trust of this population in France, providing political and economic reforms along with use of psychological methods. This aspect emphasizes the civilian-psychological-military character of the warfare that was not congruent with the professional training of the warfare system.
2. Crushing the revolt. And especially, destroying, or at least neutralizing the principal units of the enemy.
3. The rise of the status of the intelligence services and its prime relevance, not only in carrying out the military-operational mission but also for the civilian-psychological mission. The intelligence service fulfilled a central role: Collecting and evaluating intelligence material, and its use, were of primary importance, especially in the type of new war that was being carried on in the midst of a civilian population. Intelligence gathering was not only the concern of the designated intelligence bodies, or of the special units alone, but rather of all units, from the highest level to the field level that had contact with the population, the main source of information.[7]

Regarding objectives of war in Algeria, there were also voices calling for maintaining the peace and humanitarian assistance. These goals had not as yet been shaped into a security doctrine that would point to a change in military model leading to a post-modern model. The declarative and propaganda dimensions of these missions were directed towards mobilization in the political campaign to enlist French society support, and especially towards psychological warfare against the rebels. In any case, the conflict against the Algerian rebels was perceived as defending the national interests of the French homeland.

Dominant Army Profession: Warfare Commander–Technical Manager

> *A mixture of technical professionalism alongside warfare management with great authority granted to the senior military command.*[8]

Each army force had a unique military role: The Foreign Legion, the ground forces, the navy, the air force and the gendarmerie acted cooperatively in the development of the conflict against the Algerian rebels. Interforce cooperation existed on the highest level (the operational arena of Algeria), but only rarely did it seep down to mean real joint operations or common missions, with the exception of specific short range operations. The navy fulfilled a central role as it set a northern barrier that completed the separation barriers that blocked the borders with Tunisia and Morocco, preventing the smuggling of arms to the Algerian Army of National Liberation (A.L.N.).[9]

Expressions of the authority of the senior military command were striking considering the fact that the army actually ruled in Algeria and, at times, attributed ignorance to the political echelon about what was really happening in the field. In addition to its military roles, the army administered all of the relations on the civilian level and maintained contacts with the local population on a variety of issues. It supervised economic development, medical and social assistance, education and others.

The Social Attitude to the Army: Ambivalence

During these years, France was in the process of decolonization, and French society did not attribute importance to imperialist interests, especially after the Second World War. French public opinion supported the military establishment and army service was perceived as an integral part of the dominant ethos of civil service, which provided its prestige.

The war with Algeria created cracks in this public position. The army sometimes suffered from a low image in the French public, and a lack of sympathy developed for its methods. At the beginning, the war sowed feelings of bitterness and distrust in the army establishment among intellectual circles, although public opinion was not as yet sufficiently aware of what was going on in Algeria.[10] Public opinion was only clearly aroused as to what was going on in Algeria from 1956, while protests against torture and cruelty used against Algerian rebels began in 1957. The conflict with Algeria became a dominant subject on the public agenda. Along with expressions of support in French society for its military efforts in Algeria, a wide-ranging protest movement developed that now included groups on the right, on the left and intellectuals on both sides. Expressions of protest and their radicalization were primarily directed against:

- torture and horrendous acts based on a systematic perception of the army regarding warfare "benefit and effectiveness."
- escalation in operational warfare of the French army against the Algerian rebels beginning in 1958.

Most of the opposition and protests against army activity in Algeria was based on a normative ethical-national attitude according to which the military is required to behave with integrity, and this ensures national pride that is democratic, humanitarian and just. In other words, the attitude of society towards the army was imbued with dynamic and concerned involvement, ranging from support to opposition.[11]

Army-Media Relations: From Integrated Media to Manipulated Media

Relations between the media and the army can be divided into two periods in the transition from mobilized media to media manipulated by the political and military echelons:

From 1954–1956

Along with criticism of army activity in Algeria in most of the French press, *LeMonde* was prominent in its mobilized stance towards the army. Both the political and the military echelons manipulated the media from the beginning of the fighting. Heavy and severe censorship limited freedom of the press, and a state of emergency was proclaimed including all of its ramification for army-media relations. Although the general media did report on horrendous acts and military behavior in Algeria at this stage of the warfare, the issue of the war did not occupy center stage in the media and was not on the front pages of the newspapers.

From 1957 Onward

Both sides in the conflict were given central coverage within the framework of psychological warfare in which each side attempted to portray the conflict as "a dirty war." The media, for its part, stressed the clashes, the torture and the violations of civil rights, while, on the front pages of the main newspapers, the behavior of the French Army was featured. Publicity included first hand accounts and detailed descriptions of horrific deeds provided by testimony by officers who had been involved in the fighting.[12]

Conscientious Objection: Permitted

During the dispute between France and Algeria it can be said that formally, each conscript drafted into compulsory service in the French army could have refused to serve in uniform (conscientious objection) and to be sent to civil service for 20 months (which was double the army service period). Civil service was considered a legitimate fulfillment of civilian responsibility and it included missions and roles in a number of non-profit organizations. However, in light of the decolonization process, the conscientious objection option was looked upon extremely negatively in the French public, except among leftist groups.

In August 1954 a group of released soldiers formed an association that refused to agree to a decision of the political echelon continuing their service in the army in light of the developments in Algeria. This was conscientious objection ("No to war with Algeria"), which demanded the right to civilian life. The movement did not win wide public support and the soldiers were sent to Algeria.[13]

Effects of French Military Model on Military Resilience

As long as conscription service quota has been stable no apparent damage to military strength. With extended duration of service in Algeria duty to a period of 28 months a damage is caused to *Significance in the Realm of Military Spirit (cluster D)* in expressions of *fear and anxiety, exhaustion etc. (D1) and in questions about the significance of the mission (D2).*

This is not an overwhelming sense, but to relations between the fighting French force with that coming from the French settlers in Algeria. As the conflict went on, tensions between these two groups increased, which reflected a growing taste of bitterness within the French fighting forces directed against the other group. Indeed, when the duration was extended for an extra month service period than that of the settlers in Algeria, and they got no more than one vacation of 23 days during their stay in Algeria—the unity of lines among the two groups weakened, and undermined the recruits' support in the French cause. Among French recruits there was suspicion that recruits from the "*Pieds Noirs*" (this is the name given the French settlers in Algeria) received preferential treatment and were placed in safer positions. "The length of the service scared people" (author's translation); their displeasure was elaborated in letters recruits sent home to their families, and aroused feelings on the unnecessity to sacrifice their lives and endanger the lives of French reserve soldiers to allow settlers hold Algeria since "there is no gratitude for this effort."[14]

As for the reserve force: their recruiting has not always frowned upon, because of removal of the families, mortgage problems, work, etc.

Except for cases in which conscientious objection interfered and caused to are serve revolt that will be elaborated further, the principle of obedience was kept mostly.

In spite of the professional differentiation between the corps also reflected in the different uniforms and specific jargon to each force, cooperation between the troops overcame the dominant military professionalism that characterizes each force.

Not only did the damage was found on none of the components of military resilience in this respect, but an increase in the *Unit Resilience (cluster B) in the ability to maintain the organic framework, in a way that either harms or maintains the identification and the internal cohesion (B1), Coping with the burden of activity that leads to fatigue (B3), and in Organization of unit belonging and identity (B4).*

Unlike previous wars, where soldiers were assembled units with different seniority, this war was about uniformity resulting from the same age groups. Uniform and being in the same pot caused blurring of social differences. The soldiers shared the same pressures, the same packages from home, the same joys, and the same difficulties:

> The small minority that occasionally felt isolated from the majority included those who postponed their induction and later joined units that later age where the average age was younger than four to six years of their own. The older were also the ones who chose to go into track officers, but those with more personal openness quickly discovered in smaller units that were stationed in remote and isolated places, the same old urge commanders experienced during the First World War in creating a kind of family nest. In paratroopers, the gap between regular and reserve forces blurred thanks to the spirit of unit brotherhood, resulting particularly form the high level of officers.[15]

That brotherhood was the result of a learning process. This process began with briefly acquaintances created on the train to Marseilles, or swift social connections formed on board to Algeria. After that there was a training period, and again, new friendships among the young soldiers of the same age, that were placed to serve in the same area. Then there was the molded effect of life together within the unit intimacy, when every soldier's safety depends on the functioning of his friend whether it is a sentry on duty at the outpost in a distant Atlas Mountains to retain the old friends, or a paratroop unit moving from place to place, where the success of the mission depends on each soldier. The command tried to keep soldiers in the same unit throughout the period of service, for it was a lesson taken from two world wars, that changing assignment damages the morale. Reassignment occurred rarely only when the unit returned to France, usually due to a serious disciplinary offense.[16]

The Effects of Social-Military Relations on Military Resilience

Although there were mixed feelings, especially among recruits (and not among the senior officers) on the goals of conflict as well as on the means by which the conflict was conducted: Most of the soldiers lined up with their friends, who were submissive, or zealous in fulfilling their duties, and became enslaved to the military-colonial ideology. Only a few, such as advanced Christians or active Communists maintained their critical approach.[17]

The public's attitude toward the army was not in top priority for senior military command.The pressure of events that concerned them directly, the senior commanders of the French army was almost completely blind to the broader aspects of reality such as the international political dimension of war or to public opinion. Their real concern in Spring of 1958 was a sense of victory coming on all fronts, which created concern of urgency: "Maybe it's the last moment where you can take advantage of a military victory".[18]

Although there were reactions from senior military personnel who expressed their protest on the atrocities and displeasure on the similarities between the behavior of French troops and soldiers of the Gestapo.[19] It seems that internalizing the impact of public opinion under loaded operative activity, far from home, in Algeria, was marginal.

As for Media-Military relations that actually also represent the dimension of military-social-public opinion: a marginal effect was found in damaging *Significance in the Realm of Military Spirit (cluster D)* in *questions about the significance of the mission (D2),* primarily on reserve and conscripted soldiers and less on permanent soldiers. Paradoxically, precisely the permanent soldiers' morale has improved as hostilities kept on stretching, while the morale of soldiers in reserve and conscripted soldiers tended to gradually loosen in spite of their late joining to war.[20]

A partial explanation can be found in the "dirty and despicable nature that the war held in articles published by various newspapers and intellectuals (except for *Le Monde* newspaper in which its coverage was with the acceptable political official line) that highlighted the horrors of war and the reporting of senior military command itself on events in the area. Reports noted how traumatic was the confluence of citizens and young French soldiers with the violence of war.[21]

Questioning of the mission was normative values basis. Soldiers saw in the mission thrown upon them the sacrilege of Christian values and the values of human civilization.[22]

To the permitted conscientious objection in the French military model—no effect was found on its military resilience. As for post conscientious objection of recruits: this was extremely rare in the conduct of combat. Indeed, there were feelings of stress and questions about the cause for the fighting and about ways of its management, yet the obedience to the military framework was not harmed.[23]

In conclusion, the Late Modern Military Model in France was found less damaging military resilience in managing the conflict in Algeria.

Warfare Strategy in the Conflict

The War Strategy: The Algerian War is divided into two phases:

- *From 1954 until the end of 1957,* a period in which the Algerian arena of the F.L.N. dictated the pace of events leading to war, and it established itself as the leading power in the Algerian struggle
- *From 1958* the military situation undergoes a dramatic change. The F.L.N. shifts to a defensive fighting mode while loosing hold of most of the arena gripes. This is also the shift in the French war strategy: from limited tactical maneuvering to a strategy with amore violent military method against the Algerians.
- *Until the end of 1957*

The Geographical Conflict

1. The conflict was marked by blurring of boundaries in the sense that the real battlefield was at the heart of the civilian population, without front lines and an enemy in uniform.[24] The basic element of the "pacification," which included "silencing Algeria" demanded from the French army a constant contact with the civilian population followed with military, psychological and humanitarian missions.
2. The conduct of major operations in the detection and destruction of the main units of the Algerian rebels were first prominent with unstable contact and close contact combat. The operational routine conducted constant chasing after

the A.L.N.'s combatants; refinement of the chase; tightening the network; isolation and detention of the rebels; and maintaining constant contact with them. The conduct of these operations was followed by the assistance of various military forces, especially the French Air Force. However, this assistance was part of a tactical response rather than an orderly strategic approach.

Man-machine: the human factor is a major factor. Beyond the psycho-political purposes of the "pacification," the army initiated contacts with the civilian population, not only for intelligence purpose, but also for taking part in the harassment and the disruption of the insurgent groups. Low level of technology use with emphasize on the human factor in fighting characterized, relatively speaking, both the French and the Algerian side. Infantry units equipped with light weapons were the primary means at the beginning of the fighting.[25]

Professionalism as opposed to non-professionalism: The French army conducted the conflict with a high professional level in intelligence, precise planning, coordination, mobility and flexible logistics while taking advantage of close contact with the local population. Beyond this asymmetry between the French and the Algerian insurgents there was a high involvement of citizens, not only in combat but also in the requisite of the "pacification" France was committed to. As for tactical maneuver: It was of little nature-fire yet with high mobility, due to geographical conditions of the conflict.

From 1958 a shift to a strategy with less low-intensity warfare properties that was made possible by formulating an operational plan at a strategic level against the F.L.N.; concentration of the French forces fighting in order to proceed systematically and eradicate one after another various sector fighting in Algeria. Effective intelligence, special high training units and high mobility means provided by the French navy and helicopters manufactured in American allowed the use of the combat system.

The geographical conflict delimited between the parties by: Establishing dividing lines along the borders. Beyond the purpose of demarcation, these lines sought to prevent the A.L.N. in Algeria to receive reinforcements in manpower and supply; thus, to harm the insurgents' morale by their isolation. Saving the maritime border was done in combination, by the Navy and Air Force, while in the Sahara the mission was carried by tours; Partition of Algeria to three areas:

1. Operational area assigned for decisive fight and crushing of the insurgents
2. Reconciliation and peace region assigned to protect Algerian European population

3. Forbidden area assigned to evacuate the population from which was placed in the camps under the control of the military

In *man-machine* relations and in *professionalism as opposed to non-professionalism* there was an escalation between the two sides. The military organization that the Algerian insurgents managed, though not changed throughout the conflict, but in time large units were built in it, which gradually had more of a regular army. The Algerians set up bases, supplies, and training professionals in Tunisia and Morocco. Armed squads, equipped in Tunisia, penetrated Algeria carrying more sophisticated and professional weapons, which included light machine guns manufactured by Germany.

To prevent these intrusions, the French Army established barbed wire on the separation fences, placed mines (not less than 3.2 million, of various types), and set patrol units. Five brigades of mounted infantry were stationed along the eastern line. The management of the fighting with penetrated cells was characterized by an increased level. No more battles contact, but direct bombardment of the reserve forces paratroop division. This was flown by helicopters, or if the weather did not permit it was mobilized by trucks, and destroyed the enemy. These operations also incorporated aerial artillery assistance.

In 1959 the French won the battle on the borders and began a mission to defeat the insurgents in Algeria. It was a regular military warfare strategy, which included many reserve forces: three divisions of the Foreign Legion, three infantry battalions of the Colonial, a battalion Marines, airborne commandos, three divisions of the 25 paratroop divisions and all the tenth paratroopers divisions.[26]

In the conflict between France and Algeria a warfare strategy was consolidated with more aspects of a low-intensity strategy at the beginning, and less of them during its development. The different method of deployment of the French troops and the lines settings defined clearly the geographical conflict; an escalation took place with an emphasis on professionalism and machine factors with more fire and mobility.

The Effects of Warfare Strategy on Military Resilience

The findings in an investigation of these effects are based on two populations in the military warfare system: the first, a unit of the Foreign Legion and the second, French soldiers. The Foreign Legion was accustomed to regular military warfare. In the war in Algeria, the warfare strategy at the beginning of the conflict was difficult to reconcile with the professional designation that justified the existence of this unit and this was found to harm its military resilience.

Regarding the group of French soldiers, their expectations for compulsory conscription were conditioned by the ideological-national link to France's efforts to maintain control of Algeria. However, the warfare strategy adopted by the French army at the beginning raised questions about the aims of the conflict and harmed military resilience of these units. The move to operational warfare strategy that was in keeping with the military designation of these units both in warfare methods and in arms and ammunition, without limiting fire power, was found to have improved resilience.

The Warfare Strategy from the Beginning of the Conflict until 1957 damaged the cluster of *Force Building (cluster A) in training frameworks (A4) and in the loss of professional expertise (A5); The Resilience of the Units (Cluster B) in creating difficulties of dealing with the burden of activity, leading to fatigue (B2); The Military Framework (cluster C) in limited ability to maintain professional military commands and procedures (C1); The Significance in the Realm of Military Corps Spirit (cluster D) in expressions of fear and anxiety, exhaustion and others (D1) and in questions raised around the significance of the mission (D2); The Significance in Terms of Operational Fitness (Cluster E) in difficulty in defining necessary fitness due to the existing tension in training for a conflict of high intensity in contrast to a low-intensity conflict (E3).*

France had fought its colonial wars using conventional warfare. However, in Algeria, the political and the military echelons understood that this conflict posed different challenges than their last warfare in Indochina and that they should avoid analogies of applying previous warfare methods in the present conflict. The army echelon reached the conclusion that this conflict was different and required professional training in guerilla tactics. To this end, a special training center, the CIPGC, was established but did not survive as it was swallowed up in the dynamics of the events, and lost its *raison d'être*. The officers in the field complained about the lack of congruence between the lengthy period of time required to train at this center and the demand for immediate solutions. In an attempt to shorten the necessary training period, the effectiveness of the training was reduced, and thus, the center lost the advantages of its expertise. Moreover, even the training that the soldiers received dealing with psychological warfare was found to be ineffective, as most of the soldiers who reached Algeria did not understand the purpose for being there or the relevancy of the training they had received in the field.

The arrival of warfare battalions in Algeria was accompanied by shock at their lack of suitable training for the field conditions and the conflict. Most of all, this feeling characterized the soldiers of the Foreign Legion who felt frustration with their lack of relevant training for field conditions. The resulting harm also was expressed in the fact that not one new recruit enlisted in the reinforced 13th battalion during a period of ten months of fighting. In addition, there was a significant rise in the number of deserters. The loss of professional expertise was accompanied by a lowering of morale among Legion soldiers as well as harm to their self image: "The Legionnaire volunteers for difficult service full of danger, while they give him stationary and secondary missions and he feels as though he has been duped and humiliated."[27]

In the Legion unit, there was a sense of continuing frustration as there was no proportion between the investment in massive military activity and the results that were achieved by these operations. "Almost all of the operations of the brigade that had been planned in advance ended in disappointment." A sense of dissatisfaction harmed the morale of the forces and undermined their abilities to deal with the daily activities that were required.[28] This erosion was particularly salient in light of the unsuitability of the size of the military forces that reached Algeria and the missions imposed on the army there. The five tasks imposed on the military in Algeria were: safeguarding life and property, curbing the revolt so that it would not spread to other areas, control over communication lines, defense of urban centers and destruction of rebel forces. Considering the great size of Algeria, carrying out these five tasks at the same time would have required large numbers of ground troops and naval and air support forces. Although the French government directed a large number of forces to Algeria, by 1957 there was a lack of forces to carry out all of these tasks. The reasons for the lack of sufficient force derive from economic constraints and considerations of the transition to a post-modern army. However, in reality, even De Gaulle, who wanted to end the Algerian problem by granting the country independence, did not succeed in widening cutbacks due to the importance of the quantity of ground troops and continuing management of the conflict in Algeria. In any case, until 1957, efforts to rapidly integrate a large number of soldiers and, simultaneously, to attempt to continue the transition from peacetime, to the structure and methods of warfare created dispersal and confusion.[29]

The tactical organization of the ground forces in Algeria at the beginning of the conflict was not suitable for the immediate tasks in the field. This would have required internal reorganization of ground force warfare units, which would have increased the burden on other units and would have led to exhaustion. Considering the chronic lack of manpower to carry out the missions imposed on the army in the field, the attempt to create and maintain an operational reserve for an offensive that would pursue and destroy the rebel forces was found to be especially difficult.[30]

The inability to deal with the losses, despite the burdens of operational activity, "led, at times, to spontaneous acts of vengeance against prisoners or wounded enemy fighters, which became an inseparable part of the instability of this guerilla war. In isolated areas, discipline collapsed and what took place negated acceptable military practice: torture, expulsion of populations by force, use of incendiary weapons, burning of fields, destroying villages, cattle theft and others.[31] The dispute, with neither front lines nor a uniform-wearing enemy, increased the chances that certain French units would do the unacceptable. This undermined faith in fighting for a just cause within the framework of Western norms and the accepted laws of warfare.[32]

A weakening of corps morale in addition to military frustration regarding "orders to open fire" found expression in attacks of "*le cafard*" (homesickness).[33] Soldiers also suffered from fear, exhaustion, depression and personal frustration. Living the brutal realities of war with its burdensome cycles of lookout duty, calls to action, and ambushes, increased the intensity of feeling among the soldiers. An evasive and unseen enemy, guerrilla attacks on French positions, ambushes by the National Liberation Army, landmines unlimited to the front, and the general atmosphere created a sense of fear in every sentry on guard duty, and in every patrol. Soldiers were always a moving target. On the other hand, the operational burden and the growing number of victims that showed that one's comrades in arms were being picked off by the enemy led to suicides and damaged The Significance in Terms of Operational Fitness due to the existing tension in training for a conflict of high intensity in contrast to a low-intensity conflict.

The damage to operational fitness was also expressed in the feelings of the Foreign Legion. The operational activity of the Foreign Legion units was far from satisfying and this did not add to confidence regarding the government's willingness to do battle with the F.L.N. with operational fitness necessary to for victory. "Like other officers who served in Indochina, I am shocked to discover everywhere the extent to which we are slow in our ability to adapt to warfare methods that are necessary for anti-guerilla warfare. The high command is not ready to admit that in this war tanks are useless, as are jet war planes . . . because they are too fast to entrap the enemy who is on foot. This type of warfare requires only one solution, infantry, infantry and more infantry. What we really need is a super-infantryman, who is light on his feet and quick, like paratroopers, who are too few among the forces in Algeria."[34]

From 1958, there was a pronounced recovery trend in all of the above mentioned clusters. There was more awareness on the part of the political and military echelons regarding the quantity of resources and the necessary training for this conflict both on the level of supplies and logistical maintenance in the field, in the improvement of fighting methods, in mobility and the division of the operational burden to the advantage of the various corps that were stationed there and all of these contributed to increased military resilience.

The French army began to use the Legion units in a more dynamic and mobile fashion, that is, to transfer them from one sector to another with greater frequency and with greater operational fitness. The training framework was strengthened and the recruits learned how to use new weaponry.[35] The improvement in training was carried out by:

> Encouragement and promotion of training experts in psychological warfare on the basis of experience acquired in France's war against the Viet Minh; a rise in the status of the training center for warfare against terrorists and guerillas; improvement in methods of warfare, in coordination between different geographical areas and in a variety of warfare tactics (including warfare in a built up area and others) and the integration and coordination of the air force as a counter force to the F.L.N.[36]

The improvement in force building contributed to the recovery of military resilience from the harm to the clusters that has been described above. A change in the warfare strategy that expressed the warfare advantage of the army led to a sense of pride "in unique and complete operational achievement." In addition to expressions of physical exhaustion, there were prominent expressions of "a sense of maximum achievement . . . and the mood of victory." Even if this achievement also involved horrific acts, it was not on the level of the individual soldier or of a group of individuals. It was seen as a method of the military command that controlled the field.[37]

Summary of the Effects of Warfare Strategy on Military Resilience

Until the end of 1957: Great Damage to military resilience.

From 1958: Recovery of military resilience.

The more France's warfare against the Algerian insurgents included features of small-scale warfare, the more damage to military resilience there was. With the transition to warfare with more features of regular army warfare, there was a recovery in military resilience.

The trend towards recovery was limited and it is not possible to follow its expressions in greater detail as the improvement in military resilience dissolved because of the great harm that resulted from relations between the political-military echelons.

Relations between the Political and the Military Echelons

Two periods can be distinguished in a description of relations between political-military echelons in the French-Algerian conflict: before and after the rise of Charles de Gaulle.

From 1954 until the appointment of de Gaulle as President of France in June 1958: Consensus between military and political echelons regarding the objectives of the conflict and the means to manage it.

The position of the army regarding the Algerian rebels correlated to that of the political echelon and stressed the need to achieve victory by eradication of all assistance and external aid to the A.L.N. This would be carried out with the help of the political echelon and with a clear demonstration of its obligations to the military effort.

In light of the political-military and the ideological-national consensus regarding the importance of Algeria to the continuing support of the geostrategic and international status of France, it can be said that the army ruled Algeria as an autonomous body. Over and above the mutual ambitions of the two echelons, certain army officers tended to see themselves as positioned above the laws of the republic. The political echelon not only did not express opposition to this tendency, but even intensified it by generally giving in to the army, by allowing population transfer by force, by granting the army the authority of a police force, and finally by accepting the violations of international law (hijacking a Moroccan plane that carried Ben-Bella, bombing the Tunisian frontier village of Saqiyat Sidi Yusuf, among others). Even when the frustration of the recruits at seeing their commanders breaking the law and taking an activist line towards the political echelon was recognized, this remained a period of cohesion and unity among the high command of the army itself.[38]

In an October 1958 speech, De Gaulle proposes a "peace of the brave" to the rebels. De Gaulle's speech about the right of self-determination for

Algeria greatly harmed the consensus between the two echelons over objectives and means.

Even at the beginning of de Gaulle's presidency most of the senior officer corps doubted his readiness to continue to hold on to French Algeria.[39] A deep crisis enveloped the army and especially the warfare system in Algeria. De Gaulle requested complete control over military activity of the army and demanded total discipline. In the army: "A calamity with the sense that the political echelon had tricked us." The speech was received with surprise by the senior echelon; the unity of the army was undermined; there was a sense of the army being seriously torn apart. The political echelon was perceived as having lied and as having deceived all sides simultaneously, sowing confusion and misunderstanding when, at the same time that de Gaulle had given the appeasing speech, he was demanding a continuation of the warfare.[40]

The Effect of Political-Military Echelon Relations on Military Resilience

Until the Appointment of De Gaulle to the Presidency in June, 1958: The consensus between the two echelons had contributed to military resilience in *Building force (cluster A)* in *the extent of fitness, training and the number of units in training (A1) and in improving the long-term planning capabilities (A2).*

There was disagreement between the two echelons in regard to sending conscripts into warfare in Algeria, fearing public opinion at home. However, the political echelon did decide to send these conscripts to the Algerian front and the military echelon requested reinforcement of special forces both in quantity and in additional training. The consensus also affected long term planning that was expressed in the encouragement of the nuclear arms program in France, deriving from a strategic and geopolitical worldview common to both echelons. The security perception regarding the strategic importance of Algeria focused on the view of North Africa as an important front in the Cold War with implications for the security of France, both in the long and short runs. Long term planning was also relevant to the political framework of cooperation with NATO.[41] An additional fact that contributed to the encouragement of unit resilience and to the army as a whole was the fact that the political echelon had given the army a free hand in managing the conflict in Algeria, which was far away from home. Professional literature relates to the French warfare against the rebels in Algeria as a classic case of continuing adaptation to low-intensity warfare while the senior military echelon, experienced in the elements of military resilience within the army framework and enjoying a laissez-faire attitude on the part of the political echelon, was completely aware of these elements[42]; that is, both echelons were aware of the problem of military resilience and made efforts to achieve it.

This was illustrated in keeping *Unit Resilience (cluster B) in maintaining an organic framework in a way that would support identity and inner cohesion (B1) and in organization of membership and unit identification (B4).*

It is important to note that resilience on this level refers to resilience among French soldiers who had arrived from France itself and not between these French soldiers and soldiers who had been conscripted among the French who had been living in Algeria.[43]

Other efforts were carried to keep the *Military Framework (cluster C)* in *encouraging the existence of internal and external organizational communication channels (commands, accepted procedures, messages, intentions) (C2);encouraging the development of systems of learning to encode and to distribute internal and external organizational information (C3),* and *in managing the tension between what is important and what is urgent in the realm of operational routine(C4.)*

These three elements stemmed from the understanding by the senior military command of the importance of the isolation of Algeria in general and of the F.L.N. in particular and preventing contacts with the outside world. In contrast, the channels of communication within the army were important for maintaining contact and carrying out operational procedures between the navy and the air force, which controlled sea routes to Algeria. Learning and distributing information between warfare units was supplied by an advanced and effective intelligence system and the establishment of a sophisticated system of barriers designed to maintain priorities on the level of the tension of operational routine.[44] The system of barriers separated the French from the Algerian soldiers and the former were no longer required to "run over the hills" in pursuit of the rebels and thus, no longer had to deal with loss of control regarding priorities in managing the warfare. This option became an initiative whose timing was determined by the military command, while simultaneously, the command was free to plan the tension between what was important and what was urgent, so that they were no longer led by the rush of events.

After October 1958, with the rise of de Gaulle to power and his declaration in 1959 of the end of French Algeria, an attempted coup d'etat was planned

Table 3.1 Effects of Warfare Strategy on French Military Resilience

Warfare strategy	*A. Building of Force*	*B. Unit Resilience*	*C. Military Framework*	*D. Significance in the Realm of Military Spirit*	*E. Significance in the Realm of Operational Fitness*
1956–57dominance to small-scale warfare	Damage to A4 andA5	Damage to B2	Damage to C1	Damage to D1 and D2	Damage to E3
From 1958–increase in large-scale warfare	Recovery*	Recovery*	Recovery*	Recovery*	Recovery*

(A4) The framework of the training.
(A5) Utilization of professional expertise.
(B2) Coping with the burden of activity which leads to fatigue.
(C1) The ability to carry out the orders and procedures of a professional army.
(D1) Expressions of fear and anxiety, exhaustion etc.
(D2)Questions about the significance of the mission.
(E3) Defining required fitness which derives from the tension which exists in preparing for a high intensity conflict in relation to a low intensity conflict.
*The trend towards recovery was limited as the improvement in military resilience dissolved due to the escalation in the relations between the political-military echelons.

by the army, which was to be carried out between 21–25 April 1961 by the army senior command.

Even if it were possible to describe the crushing blow to all components of the cluster of *military resilience of units (Cluster B) and all of the components of the cluster of military framework (Cluster C)* stemming from the split that occurred in the army as a result of the attempted coup, only a limited picture would be obtained, which would not express the intensity of the harm done to military resilience. The attempted coup constituted a rationale for especially severe harm that cannot be broken up into elements of damage to military resilience. A military coup could be interpreted not only as splitting the army command and harming its the unity, but also as posing harm to all components of resilience, by causing a complete rift between the political and the military echelons and threatening complete chaos.

Summary of the Effects of Relations between Political and Military Echelons on Military Resilience

Consensus regarding objectives and means of warfare between the political and military echelons has been found to support and reinforce military resilience. A crisis of consensus regarding objectives and means between the two echelons has been found to greatly damage military resilience.

Algeria-France: The Military Model, Warfare Strategy, Political-Military Relations and Military Resilience

The French military model during the years of conflict with Algeria had not as yet been determined as a post-modern model. It was still a model with a socio-national link in which the military-national dimension constituted an important feature; this, despite the revulsion of public opinion and the media resulting from the horrifying acts that were carried out during the conflict by the army in Algeria. It was not the lack of consensus in French public opinion, not the media and the intellectual protest and also not the fact that conscientious objection was permitted that disrupted stability in carrying out the military missions of France in Algeria. If there were signs of mental dissonance among the soldiers, these were under control. Against this backdrop, the low level of harm to military resilience by the military model can be concluded.

Although the French perception of threat ranged from invasion of the homeland to atomic war, and the conflict with Algeria apparently was not within this range, this conflict was perceived as a threat to the homeland and its assets. France did not enter this warfare to impose peace between different ethnic movements in the same state and/or to grant Algeria humanitarian aid, steps that would have better suited a perception of threat of the post-modern model. France viewed the continuing rule of Algeria as ensuring its strategic national assets. The force structure, the central mission, the dominant military profession, all maintain a national-military link with less involvement of civilian elements and other aspects of post-modern professionalism.

The connection between warfare strategy and military resilience is significant. Warfare management with maximum properties of small scale conflict resulted in great damage to military resilience. With the adoption of more regular military warfare, recovery of military resilience could be discerned, if it had not been for the effect of political-military echelon relations leading to great harm.

The effect of relations between the political and the military echelons on military resilience undermined and damaged every possible aspect of force building. The army was split into two camps. In one, there were expressions of active conscientious objection along the entire line of command, refusal to obey orders and to go out on operations: "No one wants to be the last death of an unnamed war."[45] This camp also wanted to involve public opinion at home regarding the management of the warfare in the field. In the other camp, there were expressions of complete loss of control and the initiation of a coup d'etat by officers who banded together to overthrow the government of France.

The damage caused by military model and warfare strategy to military resilience was curbed for as long as there was consensus between the two echelons, which enabled cooperation between them:

At the beginning of the conflict, the political echelon, aided by the military echelon, was aware of the problem of military resilience in a conflict with the Algerian rebels. It certainly identified all of the links that could weaken this resilience, including public opinion and the media and their effects on the front. Thus, both echelons adopted coping strategies, each in its own way:

- Distinguishing between the Algerian rebels, who were termed criminals and lawbreakers, and the citizens of Algeria. The rebels were presented as causing damage to the civilian population of Algeria, who were loyal and who identified with France. The military objectives were meant to defeat the rebels and to return life to its normal course.
- Distinguishing between the legitimacy of the rebels' struggle and the unacceptable means they had adopted that demanded a military solution.

- Distancing the media from operational activity at the front and instituting the heaviest censorship against the backdrop of a forced state of emergency.

These strategies supported military resilience with Algeria for as long as there was relative consensus between the political and the military echelons. The use of these strategies moderated the damage to resilience caused by the military model and the warfare strategy and their success demanded the full cooperation of the two echelons. However, after the split that occurred between the political and the military echelons and within the military echelon itself, severe harm to military resilience was unavoidable.

Table 3.2 France-Algeria 1954–1962

Military Model, Warfare Strategy, Politico-military Relations	*Description*	*Military Resilience*
Military Model	Late modern	Low damage
Warfare Strategy	1954–1957 dominance to small-scale warfare	High damage
	from 1958, increase in large-scale warfare	Resilience recovery*
Politico-military relations	1954–1958 consenus on aims and means	Support in resilience**
	1958 nonconsensus on aims and means	High damage

*The trend toward recovery was limited as the improvement in the military resilience dissolved due to the escalation in the relations between the political-military echelons.

**There is no contradiction between the results of the three independent variables on the military resilience. The aspects of damage to the military resilience is clearly referred to in the military model and warfare strategy. The consensus between the two ranks reflected cohesion and unity among the high command of the army and compensated in reinforcing other aspects of the clusters of resilience.

Chapter 4

Britain-Ireland 1969–1994

Military Model

In a survey of military models, a significant fact about Great Britain is that, in 1963, it was the first Western country to cancel universal conscription and move to the model of a professional army. Thus, in the British-Irish conflict discussed here, reference is to a postmodern professional army when considering most of the aspects in regard to this model. The relations between society and the military, and relations between the media and the military still belong to the late modern model, especially in routine times. However, during the conflict being discussed, these relations were closer to the postmodern model.[1]

Perception of Threat—Between nuclear war and sub-national crisis

The fact that Great Britain adopted a post-modern model before the end of the Cold War in most of this model's components points to the fact that perception of threat created an image of relative calm in which there was no fear of danger to British security. Although the Cold War posed a nuclear threat, Britain was still satisfied with a small national defensive army as a response to a possible security threat, and at the same time, a colonial army spread throughout its empire.[2] In any case, at the time of the conflict, a more realistic perception of threat was the internal threat that undermined the sense of stability.

Due to the delimitation of the conflict years in this chapter that is beyond 1990, the perception of threat is also affected by the situation in the international arena after the end of the Cold War. At the same time that there was a relaxation regarding the real threat of Soviet nuclear power, there was the fear of loss of the sense of relative stability and an end to the status quo that had been achieved towards the end of the Cold War. An uncertain environment was created to which every ethnic or radical fundamentalist group had access.

Force Structure—A small professional army

Except for the periods of the two world wars, the British preferred a small defensive army as a response to a possible security threat, with the addition of a colonial army spread throughout the British empire.[3] During the period discussed here the size of the British army had begun to shrink despite the challenges at home and abroad.

Central Mission—Assistance to allies, maintaining international peace and humanitarian missions

During the period of the Northern Ireland dispute the declared mission of the army was to achieve law and order in Northern Ireland and to resolve the internal and external security challenges of the dependent territories. Additionally, the army could have been required to defend the country and its allies against any external threat and to intervene in many peacekeeping operations throughout the world. This was suited to the structure of a very mobile army prepared for missions such as these, as well as missions against terrorism in Great Britain itself.

Dominant Military Profession—Between the officer or the "technical" soldier to the "managerial" and the "diplomatic" officer

On a local level, the conditions of the dispute with Ireland not only required warfare qualifications and soldiers' technical military skills, but also diplomatic political abilities that were necessary for contacts with the various factions, considering the presence of the army among a civilian population. In addition to warfare ability and military deployment suitable for the defeat of the terrorist organizations and the IRA, direct British rule that was instituted in Ireland in 1972 also necessitated the ability to manage political contacts and deal with bureaucracy, to maintain relations with the population, to direct media policy management and to coordinate among the various organizations. Thus, the role of "political officer" became relevant due to the involvement of senior officers in political discussions relevant to security policy.

The Attitude of the Population to the Army—Between support and ambivalence

Marginality and centrality have characterized the British army and its relations with British society. The status of the army became more central only during the two world wars. On the other hand, during periods when the British population did not fear a direct threat to its existence, the army was perceived as "a necessary evil," a marginal profession or a temporary burden. The relations between the military establishment and society reflect long-standing agreement with other central institutions—the queen, the nobility, Parliament and the church. However, trust between the army and British society has had its ups and downs through the years. In the years after the Second World War, relations of trust had strengthened among the public, especially thanks to its proven professionalism. But in the 1960s, liberal winds were blowing through Britain bringing about the students uprisings. The operational failures in Northern Ireland, the exaggerated emphasis on "the daily dirty work" and the inflexible framework of the professional army eventually led to a public view of the army as an institution that represented authoritarian and tyrannical values. This tendency strengthened the voices in society calling for increased civil supervision of the army.

In any case, during the period studied here, the army was still perceived as an institution that safeguarded positive traditional values and that reflected the moral priorities of British class society. The army even gained symbolic legitimacy as it was one of the institutions associated with the royal family, which was proof of its international reputation and prestige.

Due to the crisis in Northern Ireland, the public supported the expressed goals of the warfare. In any event, the attitude of society to the army in its British context was one of respect for an important institution representing an inseparable and vital part of the ruling establishment and of the social fabric. Despite the information about the violence used by the army during bloody acts against the civilian population, public criticism was directed at the methods used by the army rather than at its objectives. Public opinion mostly felt that, at the bottom line, the army was doing "good work." Public support of the army also suited the spirit of reports in the written and broadcast media.[4]

Army-Media Relations—Between integration and manipulation

Intensive media coverage of events in Belfast accelerated the decision of the political echelon to send British forces to Northern Ireland in order to assist the Catholic population.[5] Until Bloody Sunday, in January 1972, the media expressed support for the presence of British forces in Ireland. The events of Bloody Sunday created a deep crisis between the army and the media, as the latter expressed an especially negative view of the warfare approach of the army. However, this crisis disappeared with the escalation on the part of IRA forces.[6]

The attitude of the media to what was going on in Ireland was selective, and at its foundation there was full support for the stated military objectives that were described as normative, encouraging both the democratization process and a return to order in Northern Ireland. Coverage of the events and condemnation of the reports about victims, which were even sometimes described as horror stories, did not turn into a negation of the justice of the political echelon's policies towards Northern Ireland.[7]

It should be noted that the written and broadcast media on both sides behaved more like political actors than like uninvolved observers. The Irish terrorist organizations had massive access to their own local media that in contrast to other media sources, did not present them as a "faltering" organization whose voice was broadcast only by others. To a certain extent, this influenced the "moderation" of the media channels on the other side, that is, the British, in the sense that they tended to less radical coverage against the British government. Thus, it cannot be said that the British media attitude to what was going on in Ireland bordered on hostility. There are those who feel that most written and broadcast news media served the interests of Britain, and were used as a channel of government propaganda, a kind of mobilized media, although not actually by conscious decision but rather by pressure imposed on it to support the policies of the government. This pressure operated on several levels: the economic context of creating communications, indirect censorship by intimidation and threat of legal action, direct censorship, and finally, self-censorship, especially among senior managers of a number of news agencies. Here the British government made an effort to integrate and to involve the state media in its battle against terrorism.

In the period between 1969 and 1993, it can be said that the media lost its reputation for independence and fairness. However, in September 1994, censorship was ended.

Conscientious Objection—Permitted as both the draft and compulsory civil service had been abolished

Due to the cancellation of the compulsory military service at the time of the conflict—the concept of conscientious objection had weakened and in its place there were other expressions. In the military realities of Ireland, there were more often refusals to participate in night operations due to fear and anxiety, feelings that characterized the first years of fighting, worry about the family at home left with no source of support and/or expressions involving pangs of conscience at the violence that cost the lives of innocent civilians. Conscientious objection was almost non-existent. On the other hand, early departures from the army began to increase due to the gap between the changes in force structure and the advanced technological system. In other words, early departures stemmed from the constraints and the implications implicit in the move to a post-modern model.

The Effect of the Military Model on Military Resilience

The events in Belfast in 1969 put the perception of threat, which had become post-modern only from 1963 with the cancellation of compulsory military service to the test. On the one hand, this was a relatively limited internal-ethnic perception of threat, and on the other, army *missions* were greatly increased. These missions were:

1. To defend and safeguard the security of the British United Kingdom, its citizens and its dependent territories, even if the threat was internal. When an internal terrorist body threatens sovereignty the army may assist a local civilian network in defeating the threat.
2. Defense against any salient external threat to Britain and to its allies.
3. Advancing the security interests of Great Britain throughout the world by "maintenance of peace and stability" under the sponsorship of a United Nations mandate.

These three general missions include a wide array of secondary missions for the army as follows:

- Achieving law and order in Northern Ireland.
- Solving the internal and external security challenges of dependent territories (such as the Falklands dispute and the war with Argentina).
- Participating in high intensity warfare as a result of an attack on NATO.
- Participating in a limited regional war assisting a member of NATO in distress.
- Granting military support as a contribution to multinational peace keeping missions under the auspices of the United Nations.
- Providing military assistance to a civilian operational authority.[8]

As a result of threat perception that is needed in demanding military missions for a *Building of Force (Cluster A)* of limited size, military resilience was damaged. There was a continual decrease in the *scope of preparation, training and number of units in training (A1)*, a disruption *in long-term planning (A2)*, and increasing *difficulty of units to utilize resources and force building capability to the fullest (A3)*. Damage was caused to the *Significance in the Realm of Corps Spirit (Cluster D) in difficulty in finding candidates for officer training (D3)*.

The transition in Britain to a professional army and the cancellation of compulsory military service caused damage to military resilience against the backdrop of the warfare in Northern Ireland. During the transition to a professional army the British assumed that they would succeed in maintaining the necessary number of service volunteers, but they were not successful. The mobilization policy ignored all types of selectivity and enlisted any volunteer who met health demands and was physically fit.

In the 1960s and the 1970s the army was "too stretched out" at the very time that it was integrating advanced technological systems that demanded long-term complex professional training and budget allocation. These requirements made it especially difficult for the battalions that carried out continuous security missions and were constantly short of manpower. As a response to the lack of manpower, the army commanders decided to lower the standards for service candidates in terms of age, education and fitness. Simultaneously, the army set up a network of schools to close the educational gaps and to raise the level of uneducated recruits. In addition, the army established boarding schools for military training for youths of 15 and over and even established complete units, especially in the navy, which was completely made up of young men who were employed in administrative and service tasks. The policy of lowering the age of recruitment increased the potential for mobilization, but it also led to a rise in early departure from the army, to high costs of training and to an extension of the training period before going on missions abroad.[9] In fact, close to 50 percent of the service list for low echelon officers was not filled by actual officers.[10]

Regarding the effects of the attitudes of society and of the media on the resilience of the army: Beyond the sense of "discomfort" of the British army in Northern Ireland, as a result of the rather selective criticism it encountered and the sense of two spheres nourished by one another,[11] no damage to military resilience has been identified. During those years, the army succeeded in breaking into the headlines of the newspapers and into the public consciousness thanks to its success in safeguarding British interests in Northern Ireland.

Summary of the Effects of the British Military Model on Military Resilience

The post-modern military model that characterized the British army in Northern Ireland led to high damage to military resilience. The possibility should not be precluded that the proximity between the year that compulsory service was abandoned and the transition to a post-modern army took place, and the year during which the clashes broke out in Northern Ireland meant that these clashes were the result of the military changes and upheavals involved in the formal transition from model to model. Thus, low-intensity warfare management in Northern Ireland posed a unique test in regard to the army and its resilience.

The reasons for the absence of damage caused by the attitudes of British society to military resilience can be attributed to the fact that:

a. British society did not make the connection between the British struggle to safeguard other British colonies overseas and continuing to hold on to Northern Ireland. The area was perceived as a historic and integral part of Britain, sharing the same fate as the British kingdom.
b. British society was provided with information about what was happening in Northern Ireland from the media, which were anchored strongly in the information system directed by the government.

The reasons for the absence of media harm to military resilience can be attributed to:

a. The relations between the media and the military in this period moved on a scale between integration and manipulation. Put differently, this variable was not yet post-modern in character. The media was not "sought after" by the political system whose activities were not dependent on its good will. In fact, the political echelon imposed its authority on the media. We may even go so far as to say that the media, in their selective reporting, gave the army support and considered its actions just for most of the period that it was stationed in Northern Ireland, so that the media were skillfully manipulated by the political echelon.
b. The political echelon was well aware of the problem of maintaining military resilience and the continuation of its effective functioning in Northern Ireland. Knowing the sensitivity of the military system to "what they were

saying at home" it invested great effort in propaganda to strengthen the legitimacy of the British presence in Northern Ireland, presenting the army as a neutral peace keeping force, safeguarding and separating the different factions lest the situation deteriorate into extreme violence.

With the rise in violence, the political echelon expressed a strict policy towards the opposing factions and negated any legitimization of their activities. They presented these factions as law breakers against whom appropriate legal steps should be taken as war criminals.

Additionally, the objectives and defense strategies of Britain were determined as follows and the media were compelled to take part:

- Strict censorship was imposed, especially until 1994.
- The rebels were delegitimized and this was anchored in international law, a security position that was applied with appropriate military strategy, including strict censorship. Army officers were trained to play propaganda roles that enabled resilience from sensitivity to the media. These roles made it possible to directly access public opinion via the media without the intrusion of commentary of any kind.[12]

Warfare Strategy in the Conflict

The management of the British army in Northern Ireland was characterized by two stages:

- *1969–1970*—Entry of British forces into Northern Ireland, cooperation between the army and the local government in an attempt to re-impose order, enforcement activity alongside assistance, with the use of military activity only to create deterrence. Low-intensity warfare was dominant at this stage.
- *1971–1994*—Management of the conflict using regular military warfare with ups and downs in the escalating operational character of the conflict.[13]

A more factual detailing presents eight military phases from the beginning of the clashes:

1. 1967–1969: Community violence.
2. 1969–1970: Intervention by the army and a temporary calm.
3. 1970–1971: PIRA initiates an attack against security forces and the Protestants.

4. 1972–1975: Failure of the Whitehall initiatives. The conflict deepens.
5. 1975–1976: Violence and a simultaneous peace movement.
6. 1977–1981: PIRA reorganizes for long-term violence.
7. 1981–1988: Full application of "way ahead." The army adopts low profile warfare.
8. 1988–1994: PIRA tries to increase the violence and fails.[14]

First Stage—1969–1970

The geographic conflict was characterized by the blurring of boundaries, unstable contact but with close-range fighting. Before the outbreak of warfare between the sides, it should be noted that cooperation between the army and the civilian population, especially the local youth, was dominant. The period has been termed "hearts and minds" in an attempt to neutralize the association with the U.S. War in Vietnam with the presence of the army among the civilian population. The cooperation was meant to maintain the moral sensibilities of the army in its dealings with the local population. However, it "grounded" the army in its involvement on a civilian level and was characterized by a blurring of boundaries between military and civilian. "Hearts and minds" was also meant to disperse young Catholics from the streets and to encourage the formation of personal relations with the British soldiers. Additionally, it was calculated to deal with soldiers "energies" by getting them into a system of personal and intimate acquaintance. Indeed, wide cooperation was achieved with young people in clubs and in schools.

When the army entered Northern Ireland at the beginning of the conflict in 1969, it did so with convenient logistic conditions, enabling the military forces to feel comfortable. On an emotional level, as well, the military alignment took care to clarify the objectives of its military presence in Ireland and the soldiers were directed to use minimal force, neutrality, discipline, awareness and security.

In 1970, there was a change. In light of the physical proximity of the army to the civilian population, the soldiers were directed to use a heavy hand after violent clashes and overnight, the army became the enemy of all of the factions as it had difficult in identifying its opponents and due, as well, to the changing atmosphere of their contacts. The soldiers' conditions deteriorated and involved night warfare at close quarters but in darkness, in the narrow streets of an unfamiliar urban area.

Man-machine was characterized by an emphasis on the human factor:

Tactics of warfare at this stage were:

- Primarily patrols on foot to get to know the area better and in order to have close contact with the civilian population.
- Response patrols rather than prevention.
- Intelligence and control—Searching for suspects with emphasis on using open patrols.
- Dismantling explosives using unsophisticated methods at close quarters.
- The means of communication between warfare units was inappropriate for operational activity in an urban environment.

Professionalism versus Populism—Civilian involvement

Despite the wide experience with low-intensity conflicts gained by Britain in their overseas colonies, management of the Northern Ireland dispute during the first years was not professional in the military sense. As opposed to the other colonies, Northern Ireland was unique and the military options were not immediately well organized in order to achieve a resolution of the conflict. The British army was, during the first stage, called upon to fulfill a civilian role, aiding the local central government, and maintaining order. It became clear that the military struggle would have to be based on a unique doctrine for Northern Ireland, as the professional British army would have to face IRA fighters composed of groups of young people from the working class slums led by a core group council of organized leadership that met in Dublin. The popular elements of the IRA, the units, were organized into brigades composed of three battalions of hundreds of volunteer citizens. Thus, military operations became more sophisticated, employing an organized doctrine.[15]

The Confrontation Method: Asymmetry between the sides in the means of managing the warfare at the same time as tactical maneuvering with little fire, sometimes with no movement.

Facing the British Army

The means of fighting used by the IRA at this stage was based on rifles as the basic weapon and an additional range of homemade warfare means.

The orders of the political echelon to the fighting forces demanded a low military profile and thus, tactical maneuverability was limited. At the stage, the real work rested upon the police. The military patrol methods were primarily on foot as the principal means of mobility. There was limited use of fire in street fighting with citizen rebels, detentions, investigations and the like.

1971–1974

The conflict took place in a delimited geographical area, with remote control warfare and less hand to hand fighting. The machine-technological factor became more sophisticated, more professional. There was still asymmetry in the warfare method.

The army moved to warfare derived from its designation as a military, which meant involvement of all of its forces. Deployment at border passes, roadblocks between the different districts, blowing up roads that linked communities with common natural boundaries—all of these were increased and supported by activity of the Royal Air Force, the Royal Navy, intelligence, infantry and others.

Despite the fighting in areas of civilian population, the army was able to set up an appropriate structural alignment, artificial boundaries to isolate their military presence in civilian areas, both for defensive and offensive purposes. This involved the establishment of bases with guard towers and walls, constructed in civilian areas, which were equipped with bunkers and reinforced by air patrol providing flexible assistance with operational speed in built-up areas. Supervision took place from these positions, and there were guard patrols, focused elimination of suspects, interception of armed personnel from afar and others.[16]

The patrol method became more mobile and was activated as offensive and not only as defensive. More control was achieved with patrol and with the improvement in control of the situation by using maps and codes. In addition, the intelligence and control networks were improved with emphasis on secret patrols and remote control methods.

The advanced technological dimension diminished the human factor and emphasized the professional, with advanced warfare equipment that was meant to provide maximum protection for the British soldier, who, in addition to his warfare clothing was equipped with a special flak jacket and helmet. The improvement in communication when managing warfare with the use of helicopters and mobile patrol vehicles enabled operational flexibility.

Simultaneously, the IRA formed a better trained military branch, which stressed improved IRA warfare methods, made possible by aid from abroad (Lybia, Iran) and included sub-machine guns, antiaircraft missiles and others. However, during the period of conflict, the IRA forces based their fighting on this weaponry and did not use more sophisticated weapons.

The professionalization of the conflict led to the establishment of the Northern Ireland Training Advisory Team, whose function was to improve the suitability of training. The strengthening of the training system was expressed in the provision of specific logistical-tactical operational qualifications directed towards Northern Ireland: Patrol techniques for urban areas, methods of riot control, how to fire at moving targets, first aid, imprisonment and curfew procedures, knowledge of the organizational abilities and tactics of the IRA, use of special warfare methods and others. Additionally, the strategic changes in warfare included the addition of Special Air Services, which granted the warfare a more operational military character in both for training forces who were already in Northern Ireland and for their own military operations. More than anything else, it should be pointed out that at this stage the British army was administered using a specific warfare doctrine that was especially formulated for the conditions of the Northern Ireland conflict.[17]

The Effects of Warfare Strategy on Military Resilience

1968–1970

On the one hand, involvement of the army in administering civilian life in Northern Ireland sometimes bordered on intervention into the personal lives of local families and individuals, as soldiers' participated in festive events and in church marches, and spent time in discotheques and pubs alongside the locals. On the other hand, these soldiers had to deal with sporadic acts of violence. This created a sense of ambivalence between the military and the civilian aspects of their roles among the soldiers. Dealing with violence was not in line with the soldiers' orders that especially limited them from opening fire. Before being stationed, they had been commanded to maintain the slogan MIDAS Touch (minimum force, impartiality, discipline, alertness, and security).[18]

The sense of frustration at operating intensively in an area that lacked geographical boundaries caused damaged to *Force Building (Cluster A) in loss of professional expertise (A5).* This was due to the ambivalence of the functional complexity in which the British warfare soldiers had to maneuver between the military and the civilian. Regarding the methods that were applied by the army at this stage: the reactive patrol resulted in damage to *Unit Resilience (Cluster B)* in *difficulty in coping with the burden of activities that causes exhaustion (B3).*

The reactive patrol eliminated soldiers' initiative and involved the investment of huge forces in counteraction operations during day and night, which resulted in a heavy burden on the resources of the British soldiers' physical and emotional strength. Additionally, the methods of communication between soldiers that were not appropriate to the field conditions at this stage caused more harm to the ability of the soldiers to cope with operational communications and added to their exhaustion.

The foot patrol, although it could be preferable to a mobile patrol because of direct orientation and close contact with suspects, resulted in damaging *Significance in the Realm of Operational Fitness (Cluster E)* in *rise in the number of accidents and operational mishaps (E2)*, and in *having difficulty in defining what the necessary fitness was stemmed from the tension existing in training for high intensity conflict compared with low-intensity conflict that soldiers had to deal with (E3)*. The harm caused by this type of patrol was a result of the lack of training and the absence of appropriate means that were needed for patrol activity in built up urban areas such as those of Londonderry and Belfast. The soldiers were equipped with unsuitable camouflage kits and suffered from lack of basic equipment necessary for fighting in an urban area.[19]

From the testimony of soldiers who participated in these patrols, their difficulties in coping with the burden of activity become clear, and the large numbers of accidents and breakdowns that occurred because of this burden can also be seen. These soldiers maintain that the army did not show any care for their personal security and sometimes aid was not even given to those wounded in the field.[20] Another factor that added to their burden was the lack of sufficient numbers of soldiers who arrived for deployment in the first transports to Northern Ireland.[21]

The intensive operational tempo led to fatigue among the soldiers as their living quarters were located in distressed areas, and the fact that they had to man lookout points for days on end exhausted them mentally and physically. The soldiers were always exposed to sniper fire when on patrol; they faced the possibility of surprise attack at the base camp and operated under a constant rain of curses and taunts from all of the different factions. As a result, the exhaustion led to many serious mistakes. Warfare in an urban area created a heavy logistic burden on the forces who required great quantities of non-standard warfare equipment that they did not have, to quell civilian disturbances.

Damage was caused to *Military Framework (Cluster C)* in *difficulties in maintaining channels of internal and external organizational communications (C2)*. These difficulties were caused by use of radio units (Larkspur A41), which were actually an obstacle to operational warfare in the field. The difficulties of radio reception in warfare vehicles within the various neighborhoods sometimes forced the use of helicopters (Sioux helicopters) as a means of communication. If the helicopter was intercepted, soldiers had to look for public telephones for communication with other personnel! This led to an additional burden and it hampered communications on all levels. Additionally, when the army ultimately understood the communications equipment was not suitable for the urban field of warfare, and equipped the soldiers with a device that was recharged with batteries, the problems of operational burden and disruption of communication were not eliminated. The batteries lasted for only 24 hours and caused disruptions and cutoffs in the midst of operations in the field.[22]

The close warfare and fighting in an urban area led to damage *to Unit Resilience (Cluster B)* in *dealing with operational burdens leading to exhaustion (B3); Significance in the Realm of Corps Spirit (Cluster D)* in *Expressions of fear, anxiety, fatigue (D1).*

During 1969–1970 operational activity was carried out in the midst of disturbances between Protestants and Catholics. More than once, soldiers were ordered to prevent a demonstration by one of the sides in the conflict, when simultaneously, they were fighting at close quarters with groups of Catholic youth. During this type of warfare, the soldiers were pelted by stones, suffered from fire directed at them at just the same time that they were aiding civilians who had been injured. This increased fatigue among the forces that were operating in the field. Lacking appropriate protection for urban warfare in which the rioters had a clear advantage in knowing the area well, and with limitations on opening fire, the soldiers suffered from both fear and tension.[23]

1971–1994

The use of preventive patrols benefited from the experience gained by the army and were improved and traechelonnslated into detailed units of training that were appropriate to the field.[24] This enabled an improvement in *Unit Resilience (Cluster B)* in *dealing with the burden of activities leading to exhaustion (B3).* The operational- technological professionalism that was demanded from a patrol of that type, was made possible thanks to technological advances that improved *Force Building (Cluster A)* in *the scope of training courses, exercises and the number of units in training(A1)* and in *the training frameworks (A4)*

The element of training in force building was considered of the greatest importance during these years after the bitter experience gained by the army during the period at the beginning of the conflict; improvement of the preventive patrol appropriately enabled a realistic definition of goals in terms of the field, and dealt with small details of vital information (for example, preparation for a patrol that was meant to enter the homes of specific families, or visit at opening and closing hours to factories, among others).[25]

An improvement in *Military Framework (Cluster C)* in *the existence of internal and external organizational communications channels (C2)* was enabled by the use of new means of communications that were suitable for mobile warfare in an urban setting as well as for mobile units. The advanced two-way radios supplied simple and efficient communication between patrols in the field and on all levels of unit command in the area.[26]

This was also made possible by intelligence and surveillance using hidden patrols that contributed to the improvement of information flow and the advances in search methods.[27]

The professionalization of the conflict significantly strengthened the entire *Force Building (Cluster A)*.[28]

With the formulation of an organized military warfare strategy with comprehensive tactical and logistic deployment, the sense of loss of any link between training objectives and professional expertise was replaced by decisive fighting against terror organizations. The new deployment structure and the virtual-artificial boundaries that were set up returned the sense of security to the continued fighting. Although the application of training methods in operational-military methods rather than only in civilian terms created negative public resonance, the military was able to adopt changes that would respond to the public criticism against it.[29]

Summary of the Effects of Warfare Strategy on Military Resilience

Warfare management with a strategy that included elements of small-scale warfare, resulted in great damage to military resilience. The move to regular army warfare moderated this damage.

Political-Military Echelon Relations

1969–1970

Management of the conflict was accompanied only by lack of consensus regarding both the means and the type of means to use, while there was consensus about the objectives of the conflict.

1971–1994

Full consensus regarding the means and the type of means to be used, except for specific differences of opinion.

1969–1970

During the first year in Northern Ireland, the British army managed civilian affairs, at the beginning, in coordination with local governing bodies, and afterwards, as a representative of the British government, which decided to take direct control of Northern Ireland affairs. At this stage there was full consensus between the two echelons regarding the objectives of the military presence in Northern Ireland, but not regarding both the means and the type of means to administer its presence.

The army attributed its increased involvement in local civic affairs to the lack of defined direction provided by the political echelon at home. Despite this involvement, which provided it with control of civilian affairs, the army viewed warfare against guerrillas as total warfare, while it felt that the political echelon would have to adopt a position of "mobilization of all social resources to deal with the revolt." The army was split between two different political directives, one coming from the Northern Ireland government at Stormont and the other coming from the central government in London.

Despite the presence of the army in the field, the political echelon imposed most tasks on the police. Even when the British government began its direct rule of Northern Ireland, it remained split between preference for the Catholics and Protestants of Northern Ireland, and those of the political echelon at home, and tried to lower the intensity of disturbances. The senior command of the army complained about the lack of directional clarity provided by the political echelon and signaled that, considering the character of the conflict, the civilian authority should prefer military thinking. In other words, the military echelon tried to provide itself with the freedom to make decisive political decisions on the basis of operational demands. However, the political echelon, in fact, granted freedom of operational action to the police force. In the absence of direction to the military activity, the soldiers were forced to make arrests without legal authorization. Even in verbal requests to protesters and organizers of disturbances, they were forced to be restrained and "refined" or they might find themselves being found guilty by a civilian court for extreme use of force and verbal abuse.

In May 1969, it was the police who operated in the field rather than the army. The police took an active role in arrests and attacks on terrorists, with the objective of giving the conflict the character of legal struggle against lawbreakers, in line with directives from the political echelon. The army was supposed to grant "civilian assistance" only, to maintain a low profile and to let the police fulfill its tasks.

1971–1994

The orders of the political echelon were issued in coordination with the senior military echelon and included:

1. The enemy should be identified and the reasons for his/her existence clarified.
2. Resources and manpower should be coordinated at all levels of warfare alignment in the field.
3. The enemy should be overcome and weakened tactically.
4. The enemy should be isolated and frustrated by every political and military method.
5. The enemy should be destroyed.

Full control of the army in Northern Ireland coordinated with the directives of the political echelon, which operated as a regulator of military operations via political activity. This included orders to decrease or increase the warfare profile, while the two echelons agreed that a political solution was more practical than a conflict solution. Military solutions were only meant to gain time until convenient conditions were established for Britain to reach a political solution. Put differently, both the political and the military echelons viewed military activities as a means that would force the IRA to admit its defeat and announce a cease fire as an opening point for the possibility of a political solution. And that is what occurred.

The Effects of Military-Political Echelon Relations on Military Resilience

1969–1970

A sense of frustration in the army from the absence of clear political direction resulted in the expert opinion that damage was caused to *Unit Resilience* (Cluster B) in *dealing with the burden of activity that led to exhaustion (B3); to Military Framework (Cluster C)* in *limited ability to carry out orders and regulations of a professional army (C1),* and in *Significance in the Realm of Operational Fitness (Cluster E)* in *the accumulating burden of using technological methods that created dilemmas in maintaining norms of service and fitness(E1) and in a rise in the number of operational accidents and mishaps (E2)*

For the army, the events of Bloody Sunday that occurred later were an operational mishap directly resulting from the frustration that led the soldiers to fire exaggeratedly causing dozens of victims.[30] On the one hand, the police force was harmed by need to use military warfare methods that they were not used to, and on the other, the army was harmed because the political echelon granted it the role of providing assistance to law enforcement agencies instead of a role that would take advantage of its military designation.

The limitations on opening fire and the low profile conflict management that was dictated by the political echelon, the "easy finger on the trigger," the coarseness, the rude behavior and cruelty were the dominant expressions of this harm to the military. The soldiers explained their horrific acts as frustration at being "sitting ducks," as the limitations imposed upon them left them with a sense of insecurity in a hostile environment.

1971–1994

Thanks to the full consensus between the two echelons, there was no damage done to military resilience.

Summary of the Effects of Military-Political Echelons Relations on Military Resilience

The absence of a consensus in the means and type of means in managing warfare in Northern Ireland at the beginning of the conflict led to serious harm to military resilience.

Britain-Ireland: Summary of the Connection between Military Resilience and Military Model, Warfare Strategy and Military-Political Echelons Relations

During the period of the conflict, there was severe harm to the resilience of the post-modern model army of Great Britain. Although relations between society and the army and those between the media and the army were not characterized by the terms of the post-modern model, the effects of the model on military resilience were unavoidable. The fact that the year of transition to the post-modern model in Britain (1963) and the year that the disturbances in Northern Ireland broke out (1969) were so close to each other clearly emphasizes that the combination of a sensitive perception of threat and stressful military missions, which were due to the far reaching changes in the process of force building, were enough to cause great damage to military resilience.

Thus, the possible dangers involved in the transition to a post-modern army that may occur close to the time of the start of a conflict of any sort cannot be ignored, all the more so considering the damage to military resilience from such a transition while conducting a low-intensity conflict.

The unsuitable warfare strategy, which was poorly organized for fighting in an urban setting, caused great harm to military resilience. The elements of small-scale warfare that characterized the fighting, and the management of the conflict with ineffective means posed a complex challenge to resilience and weakened it. Only after the need for a change in the development and application of operational methods was understood, was the necessary training to deal with the conditions of the conflict made possible.

Despite the consensus between the military and the political echelons regarding the objectives in the conflict, the absence of such consensus about both the means and the type of means caused great harm to the resilience of the army. A possible explanation for this may be proposed considering the professionalization framework of the senior as well as the junior army command profile that characterizes the post-modern army. It appears that an officer whose professional training spans not only warfare management but also civilian-diplomatic-political dimensions that are necessary in consideration of the nature of a low-intensity conflict like that in Northern Ireland, creates tension between the two echelons even when this only means a lack of consensus in the means of warfare and their type. The professionalization of the officer in terms of being responsible for areas that are also the responsibility of the political echelon weakens the differentiation between them while the military develops political tendencies due the widening of its authority in managing warfare far from home. Evidence of this explanation can be seen in the violations, although minor, of orders issued by the political echelon on the part of the military echelon, while this did not prevent pro-British demonstrations in Ireland from circumstantial considerations.

In any case, the damage to military resilience resulting from the absence of consensus about the means that existed between the two echelons was immediately reversed with the resolution of differences of opinion between them.

Table 4.1 Britain-Ireland 1969–1974

Military Model, Political-military relations, Warfare strategy	*Description*	*Military Resilience*
Military Model	Post-Modern military	High damage
Politico-military relations	1969–1970 Consensus of aims; non-consensus of means	High damage
	1971–1994 Consensus of aims and means	Resilience Recovery
Warfare strategy	1969–1970 Small scale warfare	High damage
	1971–increase in large-scale warfare	Low damage

Chapter 5

Russia-Chechnya 1994–2000

Military Model

Russia is in the process of moving from a late modern military to a post-modern model. Russia has a unique model composed of features from a variety of models. Russian military tradition does not enable a transfer of military structure to a post-modern model in the terms discussed in this book. However, if we take the changing trends into consideration we may determine that Russia is moving from a late modern military model to a post-modern army in a process of gradual change.[1]

November 1993 represents a turning point in the perception of military model by the senior political and military leadership in light of the new geopolitical environment. One salient factor delayed the adoption of the post-modern model: the political instability, the administrative chaos and the weakened economy, all of which existed during the period examined.[2] However, the political echelon, conscious of the loss of popularity of the army in Russian society has attempted to change the military model not only for geopolitical, military and economic reasons, but also for political reasons.

The Perception of Threat—Between nuclear war and internal crisis.

After the Cold War, the perception of threat changed from viewing threat as global, mainly a struggle between super-powers, to a more complex threat from the southern periphery of the Russian Federation and from Third World states in the process of developing nuclear weapons. Russia had not yet abandoned a strategy dependant on nuclear weapons. During the period discussed, the official Russian military doctrine had proclaimed a final abolition of the defensive strategy of Gorbachev, which led to significant cutbacks in agreements to reduce conventional forces in Europe. However, as the Russian conventional military superiority had subsequently been reduced, the Russian army attributed defensive and offensive importance to nuclear weapons. This was especially due to the fact that perception of threat also included fear of nuclear weapons being used against Russia.

Regarding the perception of threat on the level of internal crisis, warfare management in a low intensity conflict constituted a significant potential dimension. The use of the army in international attempts to maintain peace, stationing troops outside of Russian national territory and conducting operations to maintain peace within the areas of the Former Soviet Union became a new component of Russian threat perception. The military doctrine determined that use would be made of army units in internal disputes in order to aid Interior Department forces of the Russian Federation in delimiting and closing off an area of dispute, in suppressing armed conflict and in separating the various factions involved in the conflict, and in protecting the strategic goals of Russia.

Threat perception and its sources, according the Russian military doctrines in the period examined here, is detects—local wars–that stem from conflict of economic, military, religious and/or ethnic interests between the republics of the former Soviet Union. According to official Russian doctrine, in local wars and conflicts, use would be made of all the available arsenal of arms, from small, light arms to the most up-to-date and sophisticated weapons and military equipment. The same applies of course to:

- Regional wars: Wars that have broken out in neighboring countries that have large armies (such as China, and Islamic states).
- Wide-ranging global wars: Wars that break out among countries possessing strategic nuclear or conventional weapons, which are able to reach Russian territory.[3]

Major Mission Definition—Defense of the homeland, in addition to recognition of support for allies, and international cooperation in operational initiatives to impose peace when these are perceived as necessary, especially in the republics of the Former Soviet Union.

The major mission definition of the army centers around creation and development of forces that will be able to defend the independence, the sovereignty and the territorial integrity of the country, civil defense, and the vital interests of society and country in line with the military-political situation in the world. In a world where the global conventional and nuclear threat has diminished, defense of the homeland is on the level of local wars that present a threat to the stability and the peace at home and abroad.[4]

Military Structure—A large army based for the most part on conscription

The army is not based on up-to-date technology. Although Russia is a nuclear power, there has been no breakthrough leading to the new generation of weaponry, as in rival countries such as the United States. During the years covered, Russia adopted a new military doctrine (R.M.A.) centering on a modern technical revolution as the prime objective. Although a policy of reduction in the size of the army and induction on a professional basis is being considered, this is being done gradually against a backdrop of opposition from a number of directions, primarily from the senior military echelons. This is not a striking feature in the conflict between Russia and Chechnya during the period being discussed.[5]

The structure of the army constitutes a practical aspect, which should be translated from the perception of threat and the central military objectives. In Russia, following the Cold War, there was difficulty in translating the perception of threat into practicality. Documentation of the military doctrine indicates a paradox within the military structure. On the one hand, local wars and armed disputes are clearly presented as the central threat to Russia. On the other hand, the operational strategic principles testify to the development of preparedness for military operations based on air force and naval capabilities on a global level. This appears to indicate the lack of governmental ability to advance defined military reforms due to a failure to achieve consensus in practically applying the new doctrine. Indeed, Yeltsin's proclamation of 1996 regarding the end of conscription and plans to diminish the size of the army were not applied during the period discussed.[6]

Dominant Military Profession—Battle management with "technical," "managerial" military leadership

The tradition of the Communist regime regarding the political-military leadership still exists. The conflict with Chechnya was conducted primarily on the basis of a battle-technical orientation as significant diplomatic authority was not accorded to the army leadership in the field. Military echelon decision-making in everything connected to contact with the population of Chechnya or its leadership was dictated solely by the political echelon.[7]

It must be noted that the army establishment had begun to develop into an autonomous body with no connection to any specific political regime, and did not avoid publicly expressing positions that were in opposition to those of the political echelon. However, its authority remained primarily in the military-technical areas only. To the extent that senior Russian military officers established contact with leaders of the rebels, their initiatives did not receive the support or recognition of the political echelon. A clear expression of this can be found in the demand of the senior army command to be given protocols of detailed orders from the political echelon for conducting the warfare in Chechnya. This was due to the deep disagreements between the two echelons that remained unsolved. Thus, the senior army command wanted to ensure in writing the legality of the instructions of the political echelon for warfare, as this legality was, in their opinion, in doubt.[8]

Relations between the Military and Society—Between indifference and support

The relations between Russian society and the army ranged from indifference to its institutions and its symbols to worry about the soldiers fighting in a war that, at the beginning, was perceived as unjustified.

The absence of a national-military ethos characterized the atmosphere of relations between the army and society against the backdrop of lack of motivation to serve in the army, indifference to military obligations and to the defense of both the homeland and the nation among a large section of Russian youth. At the base of Yeltsin's initiative in proposing an end to the draft was the unpopularity of the army and the decline in respect for it. Russian involvement in Chechnya deepened Russian society's sense of revulsion for the army and increased the opposition to army service.[9]

Regarding the attitude of society to the Russian presence in Chechnya: it can be described as tending to rise and fall in line with what was happening militarily.

At the Beginning of the Conflict: There was strong opposition to the military incursion and clear delegitimization. This became sharper after the military failure of the first campaign. The public opposed both the objectives of the conflict and the military solution. In a survey conducted in April 1995, 70 percent to 80 percent of Russian citizens thought that Chechnya should not be forced to belong to Russia. The public did not blame the army but rather, the political leadership. The image of the army was that it "was doing what was decided by the political echelon—carrying out orders according to the law." However, the public did not excuse the army for its part in the situation, as it had failed to stop the politicians who were leading the war.[10]

The anti-war movement, which developed in opposition to the war in Chechnya, should be noted. Although it was not as yet anchored in civil culture after the fall of Communism, it succeeded in gaining influence, calling for improvement of general conditions for the fighting soldiers and compensation, as well as improvement in sanitation and medical care. For soldiers who committed offenses in the line of duty, it demanded that they be brought to trial.[11]

From 1999—The beginning of the second campaign: There was increased legitimization in public opinion, although it was based on a manipulative propaganda campaign on the part of the political leadership. Here, too, it may be assumed that the result of the Russian failure in warfare, and the deaths of many Russian soldiers, as well as the increase in terror activity against civilian targets in Russia created an atmosphere that justified public support for governmental policy. Chechen acts of terror throughout Russia affected public opinion, and especially the bombing of multi-storied apartment buildings, which left many victims. In addition, the Chechen incursion into neighboring Dagestan contributed to a change in Russian public opinion about the Chechen problem. The Russian government exploited these events to the limit in order to enlist public support for the justification of a military operation that would lead to a solution of the Chechen problem.[12]

Relations between the Army and the Media—Manipulated.

At the beginning of the conflict: The relations between the army and the media were similar to those of society to the army, although the media demonstrated strong opposition and hostility at first; and as the war continued, this became support, resulting from political and military echelon manipulation.

During the first campaign, the media stressed its opposition to a military solution to the conflict and urged the leadership to open negotiations with the leaders of Chechnya. There was media coverage of the failures of the military operation in addition to the sharp criticism of the military establishment for agreeing to carry out the orders of the political echelon, and for hiding the details of the operational failures, and in general, for lack of support for the soldiers and the failure to publish complete casualty lists. The soldiers were described by the media as those who were carrying out their responsibilities under difficult physical and psychological conditions.[13] Soldiers who had been killed in the war were described as having been led astray, as the media also attributed a central role to human rights.

From 1999—The media supported military action due to the manipulative pressure of the political echelon that heavily and strictly censored them and prevented any contact between "unsupervised" media representatives and forces fighting in Chechnya. In order to improve control using the media, a presidential decree was issued ordering the establishment of a Russian Information Center. Its most important function was to supervise and to filter reports from the battlefield before their distribution by the Russian media and to supervise information publicized in Russia from foreign media and information sources. The media was cooperative and adopted the terminology and the jargon of the military that was meant to "soften" the messages to the public from the battlefield. Surprisingly, most of the Russian journalists, with a few exceptions, cooperated with the establishment and did not complain about the media supervision.[14]

Conscientious Objection—Forbidden by law

Army service was based on conscription. However, it should be noted that, during the war in Chechnya, regular army soldiers were permitted to refuse to participate in the fighting in Chechnya, and service there was voluntary. In other words, although a volunteer army had not as yet been established formally, the general rule permitted fighting in Chechnya on a voluntary basis.[15]

The Effect of the Military Model on Military Resilience

In an analysis of damage to resilience it should be remembered that, during this period, the Russian army was involved in a continual dynamic consisting of three different areas:

- Reorganizing its units from the pattern that had been dictated by the Communist regime, and that had undergone changes as early as the Gorbachev period.
- Redesigning relations between the political and military echelons
- Creating a new doctrine of security directed towards military revolution affairs (R.M.A.), which included a change in the military model.[16]

Therefore, the effects of this dynamic on military resilience cannot be ignored. This dynamic in effect created structural and functional conditions for especially vulnerable military resilience. Indeed, the senior military command was most aware of the vulnerability of the military at this time. The internal army problems, the policy of cutbacks in appropriations, the damage to the living standards of the fighters were all results of the transformation that the army was experiencing during the period covered.

At the Beginning of the Conflict

The critical, often hostile media were an important factor in the failure of the Russian security forces in its campaign to achieve public support for the war in Chechnya. The link between the media and public opinion at the beginning of the conflict damaged discipline and morale among the soldiers. Reports of the extent of the damage to Russian forces and the high casualty toll as well as the impact of these reports on public opinion brought about damage to *Significance in the Realm of Military Spirit (Cluster D)* in *Expressions of fear and anxiety, exhaustion etc (D1)* and in *questions about the significance of the mission (D2).*

In the Second Campaign From 1999—The attitude of society to the army and the approach of the media to the military were dictated by manipulation of the political echelon that was aware of their importance for the morale of the fighters. Indeed, during this period, no damage was found to have been done to the resilience of the military as a result of media coverage of the war or resulting from public opinion at home.

Summary of the Effects of the Military Model on Military Resilience

The Russian army, which, during this period, was modeled between a late modern and a post-modern army, was found to be minimally damaged in its resilience. Except for two variables of the military model, that is, relations between army and society, and relations between media and the military, no connection was found between the other variables of model and army resilience.

It should not be concluded from these findings that military resilience was not in crisis. Methodologically, in the causal link between the military model and army resilience, two important considerations are involved:

1. Factors that derive from complex circumstances in the three areas discussed above, reorganization, the creation of new relations between the military and the political echelons, and the creation of a new security doctrine, and not as a result of specific characteristics of the military model.
2. A factor linked to the relations between political-military echelons, which will be further discussed, whose damage to military resilience weakens the causal link between military model and military resilience.

The Warfare Strategy

Regular conventional military combat with marginal features of small-scale combat. Use of unconventional weaponry during the development of the Grozny campaign

In each of the Russian military campaigns against the Chechens, the combat strategy was based on dominance of regular military combat, in addition to marginal features of small-scale combat. The difference in fighting between the two campaigns can be discerned in the tendency towards military escalation and the use of unconventional weapons. During the war, all limitations on the military were removed.

The First Chechen War 1994–1996

The Geographical Conflict—Fighting in an urban area, unclear boundaries, fluidity in contact and fighting at close quarters during the battle.

The entry of Russian forces into Grozny placed the Russians in a challenging situation from the beginning: *fighting in a built-up area and unclear boundaries between fighting forces.*

The use of Russian tanks to capture Grozny met with opposition by Chechen forces, who moved towards them in hiding, using side streets, groups of buildings, and probably through the underground sewer pipes and telephone network channels.[17] The Chechen method was to drive out innocent-looking civilian cars from side streets, behind a line of Russian tanks, and while in motion, to fire an RPG rocket against the last armored vehicle in the line and then to immediately disappear into another side street before the Russians had time to react. At this time, other Chechen fighters were attacking the first tank in the line and snipers were firing at officers, signal operators and Russian soldiers. This is what is meant by *fluid contact and fighting at close quarters.*

When the battle for Grozny was renewed, Russian forces applied lessons about fighting in a built-up area, but the geographical features of the conflict remained as they had been. The battles took place in three dimensions: Both the Chechens and the Russians activated underground forces, who moved secretly through pipes of the city and who burst out to attack the rear of the enemy. Even underground battles took place. The Chechens were especially expert at triple tier attacks, which became an extremely effective tactic. The attacks took place at the same time using three forces, one which would be moving underground, the second which would be at ground level and the third which would be on the upper floors of the buildings, while all units were coordinated and each had its own mission.[18]

The Second Chechen War 1999–2000

Fighting was broadened beyond built-up areas, according to the Russian attack plan.[19] Afterwards, fighting again focused on capturing Grozny and clearing it of Chechen troops.

Close-range fighting was still taking place here and there,[20] but the Russians made an effort to maintain long distance fighting with the "escalation method" in Grozny, in an attempt to suit the diverse weaponry to the various areas. Fighting focused on systematic destruction of Chechen objectives and infrastructure in Grozny, while land forces gradually occupied a number of the city suburbs, without trying to penetrate the center of Grozny and to fight with the Chechens at close quarters. Artillery fire was employed from the early stages of the attack in order to prevent Chechen forces from establishing close contact fighting ("hugging tactics") or attaching themselves to the Russian forces, and thus, lessening the advantage provided by artillery and Russian air support.

Man-machine

Characterized by combining the two against a backdrop of massive use of Russian military tactics, which had already unleashed a huge amount of fire power during the first campaign. As the campaign continued, Russia activated unconventional weapons and the human factor became marginal.

The Russians shelled Grozny heavily and advanced with the support of infantry, tanks and armored vehicles. However, there were Russian combat failures on the level of command and control, coordination, preparedness and training, and additional military errors along with sophisticated combat tactics used by the Chechens, who exploited these failures. This led to the involvement of the *human factor* in the fighting. As mentioned, from an early stage, all limitations had been lifted from the Russian army. The fire power used against Grozny in January-February 1995 was on a level that had not been seen since World War II.

The Second Chechen War

In addition to massive fire power, large numbers of ground and air forces were used, including unconventional fuel-air explosives, which had also been used on a limited basis during the first campaign. In line with the lessons of the first war, the Russian military command requested confirmation from the political echelon to make use of unconventional weaponry during the second war. The political leadership agreed to the use of thermobaric weapons and fuel-air explosives with a number of restrictions. It should be noted that the fire power of fuel-air explosives is greater by 1.5–2 than the explosive power of conventional explosives. Despite the amount of fire power used, the human factor existed as well, although it was marginal. This was forced on the Russian forces when no adequate solution was found to the use of the "hugging technique" or "close contact" employed by the Chechens, preventing the Russians from efficient use of their preferred artillery and air fire power.[21]

Professionalism versus Popular Struggle / Citizen Involvement

Professionalism characterized Russian combat while the Chechens used tactics of a popular struggle, although there were also professional elements as the Chechen fighters had served in the Russian army in the past. The popular character of Chechen combat in the two wars derived from their inferiority against the power of Russian forces rather than Chechen lack of military professionalism, professionalism that was expressed in their operational planning.

Both in the first and in the second wars, the Chechens adopted the "hugging technique," keeping their forces as close as possible to the Russians in order to reduce their vulnerability to Russian artillery and the Russian air force as much as possible. *During the first war*, most of the fighting was done by an organized group of fighters under local leadership. *During the second war*, a massive number of civilians was mobilized by Chechen fighters.[22]

Coping Methods

There was asymmetry between the sides in managing combat, simultaneous strategic maneuvers with strong fire power and movement on the part of the Russians.

With the exception of the element of asymmetry in management of the war, as stated, in both wars Russia used all of its military power that led to escalation, a result of the lessons learned from campaign to campaign. Regarding Russian troop movement, it was more limited during the first war and focused on the city of Grozny. However, the second war required broader movement of forces, dictated by the plan of attack that included: advance of Russian forces to the Tarak River and clearing the area of Chechen forces, advancing to the suburbs of the capital, Grozny, occupying the lower parts of the Caucasus Mountains that dominate the city, encirclement of Grozny, and capturing the city and clearing it of Chechen forces.[23]

The combat strategy was integrated with small-scale warfare and regular army combat, escalating in the severity of the military methods adopted, which were unconventional. The geographical area and the asymmetry in the combat methods were the salient features of small-scale combat.

The Effects of the Warfare Strategy on Military Resilience

The salience of great damage to military resilience can be discerned in the first campaign of the war, between 1994 and1996. This damage became significantly moderate during the second campaign with lessons derived, and against the backdrop the broadening of combat strategy to regular military strategy on all levels. The following are the findings regarding the effects of Russian combat strategy on military resilience.

The damage to military resilience was in the cluster of Force Building (Cluster A) in The extent of preparedness, training and the number of units training(A1), The ability of the units to utilize resources and capabilities in order to build force (A2), and in Utilization of professional expertise (A3); Unit Resilience (Cluster B)in Dealing with the burden of activity causing exhaustion (B2); Importance of the Realm of Corps Spirit (Cluster D) in Expressions of fear and anxiety, and exhaustion, among others (D1), Questioning the significance of the mission (D2), Recruiting candidates for officers' duties (D3); In the Military Framework (Cluster C) in The existence of internal and external organizational communication channels (directives, regulations, messages, intentions) (C2), The existence of learning mechanisms for structuring, encoding and distribution of knowledge both internally and outside of the organization (C3); in the Importance of Operational Fitness (Cluster E) in Activation of technical means that creates dilemmas for maintaining norms of service and fitness, that is, use of technological means in a situation that is not in keeping with norms of service and fitness to operate accordingly (E1), The number of accidents and operational mishaps (E2), Defining the required fitness deriving from the existing tension between preparedness for the high intensity conflict in relation to that of a low intensity conflict (E3).

The decrease in the use of resources by units was expressed in the failure of the system of recruitment and personnel management. Most units were manned by fewer soldiers than required and thus, fighters were transferred from unit to unit, in order to fill the lines. Therefore, in actual fact, "improvised units," disunited and untrained, formed the military framework in Chechnya.[24]

The features of the geographical conflict were the Russian's weak points, as their combat plan was based on methods (the use of tanks and armored vehicles) whose flexibility, maneuverability and fire power were limited in built-up areas (due to difficulties of visibility, elevated angles etc.).[25]

The inability to exploit resources and the loss of professional expertise (even though this was especially limited and replete with problems and failures from the beginning of combat) were prominent considering the Chechen familiarity with the field of combat. When Russian forces attempted to conduct infantry battles, the infantry equipment, which included clothing, and defensive flak jackets, were not utilized to the fullest because their weight was not suited to field conditions and limited the soldiers' maneuverability and movement, making the fighting more difficult. In addition, the Russian infantry was mechanized for battle in open areas, but lacked the training for fighting in urban settings or mountainous areas, and which demanded fighting on foot rather than mechanized battle. This caused more severe damage during the first campaign despite the fortifications that the Russians managed to set up. The Russian weak points remained their passive deployment (so that, more than controlling the city, they controlled a collection of positions and bases with no communication between them and with no possibility of mutual support), remaining at their bases and their vehicles and avoiding movement on foot. Therefore, they could not find substitutes for main traffic arteries if these were blocked. So the Chechens succeeded in neutralizing the Russian advantages, exploiting their weak points that included passivity, division, relying on main roads, abstaining from movement and fighting on foot. This had the effect of isolating Russian positions and military posts, turning the Russians into a collection of divided forces instead of a large concentrated force and preventing them from making use of their power. *This damage led to additional damage to Unit Resilience (B2).*

In light of the serious problems of command and control, especially in coordination between the army and the forces of the Interior Ministry, *damage was created in the Importance of the Realm of Corps Spirit (D1, D2, D3).*

The Chechen combat advantages in a built-up area and the exposure of the Russian forces to the maneuverability of the Chechens in the urban maze created the feeling among the Russian soldiers that they were "sitting ducks." Their limited maneuverability, the Chechen success in isolating the various positions, the channeling of the Russians to vulnerable areas, and the exploitation of the urban environment to attack from areas that were shielded from Russian tank fire- all of these created a feeling of fear. The preparedness that was necessary for this kind of fighting and the operational limitations that it dictated exhausted the forces in the field and pointed to their powerlessness. At times, the Chechen fighters wounded a Russian soldier on purpose and then wounded those who came to his aid. This method contributed to the undermining of Russian morale and Russian demoralization. Early in one of the battles during which the Russians were surrounded and being fired upon, one of the Russian commanders shouted, "They are firing on us from every direction. Where is our aid? The entire battalion is wounded and lying in the station. Think about how to evacuate the wounded! We don't have any vehicles left. Do you understand? Work on evacuating the wounded! We are all dying. We need help. Do you understand?"[26]

Just before the planning of the attack of August, 1996, in Grozny there were hardly any officers who had participated in the March attack. The rotation of Russian officers that was meant to provide the largest possible number of officers with experience in the field, determined that officers would remain in Chechnya for only two months and not for a year, as required by regular soldiers. This rotation caused many officers to become passive and uncaring. Many of them were not convinced of the importance of the mission and only wanted to complete their short period of service in Chechnya without complications. Thus, many of the officers in Grozny avoided any activity that was not necessary. This rotation also damaged *Military Framework (D2, D3) and Importance of Operational Fitness (E1, E2, E3)* The frequent rotation did not enable forces in Grozny to develop an "organizational memory." In the absence of overlap, any lesson that officers had learned in the March attack had been lost.[27]

The character of the fighting, which included elements of the three different levels of combat (small-scale, conventional and unconventional), became fertile ground for damage to meanings in the realm of operational readiness. On the other hand, the tricky character of Chechen combat was not in line with Russian expectations and led to a rise in the number of accidents and in operational mishaps.

There were also shortcomings in command unity: forces from different commands were attached to one another, both from the army and from internal units of the Interior Ministry, and some of the soldiers were new recruits with no experience. These deficiencies complicated the channels of communications between fighters, and giant traffic jams ensued. Even as early as the first advances into Grozny, a serious problem arose of "friendly fire" due to communication problems. Aid and evacuation forces that had come to help were fired upon by the forces that needed help.

Over and above the effects of force structure on the complicated channels of communication, deliberate damage was caused by Chechen forces who took pains to throw Russian communication into disorder. First, they used snipers against command officers and radio operators to disrupt the Russian command and control. This damaged Russian transfer of orders and messages. Second, as they knew Russian, the Chechens succeeded in disrupting transfer of messages between Russian fighters and to mislead them in managing operational communications.

The damage to the Russian ability to use technological methods and damage to their fitness resulted in the Russian shelling that destroyed buildings and buried many civilians under their ruins. The uninhibited and indiscriminate use of weapons- especially flame-throwers and white phosphorus, injured many civilians. In addition, there were Russian units who purposely targeted civilians and committed acts of robbery, rape, murder, and pre-planned execution.[28]

Summary of the Effects of Combat Strategy on Russian Military Resilience

Russian combat strategy was found to cause great damage to military resilience. Although most of the damage to resilience was prominent during the first campaign, it appears that the character of Russia's conventional combat as well as its unconventional combat both had an effect on the damage done.

Table 5.1 Effects of Warfare Strategy on Russian Military Resilience

Warfare strategy	*A. Building of Force*	*B. Unit Resilience*	*C. Military Framework*	*D. Significance in the Realm of Military Spirit*	*E. Significance in the Realm of Operational Fitness*
1994–1996 Integrated warfare (small and large)	Damage to A1, A2, A3	Damage to B2	Damage to C2, C3	Damage to D1 and D2	Damage to E1, E2, E3
1999–2000 Large scale warfare	Limited Recovery	Limited Recovery	Limited Recovery	Limited Recovery	Limited Recovery

(A1) The extent of fitness, training, and the quantity of units in training
(A2) The capability of long-term preparation
(A3) The ability to utilize resources and strength building abilities by the units
(B2) Coping with the burden of activity which leads to fatigue
(C2) Maintaining channels of internal and external organization (orders, procedures, messages, intentions)
(C3) Maintaining learning systems for encoding and distributing internal and external organizational information
(D1)Expressions of fear and anxiety, exhaustion, etc.
(D2) Questions about the significance of the mission
(E1) Activating technological methods, which create dilemmas in the realm of maintaining service norms and fitness, that is, use of technological methods when they are not congruent with the service norms, and the readiness to act accordingly
(E2) The number of accidents and operational mishaps
(E3) Defining required fitness which derives from the tension which exists in preparing for a high intensity conflict in relation to a low intensity conflict

Relations between the Political-Military Echelons

At the beginning of the warfare there was no consensus between the political and the military echelons regarding a military solution as a means of managing the conflict, both in military planning and in methods of warfare. In the army command, there was also some opposition to the objectives of the conflict. The intensity of the lack of consensus increased especially in light of the deceit of the political echelon. In the first campaign for Grozny, the lack of consensus between the echelons included disagreement about conflict objectives.

During the Second Campaign in Chechnya

There was full agreement between the two echelons regarding objectives and weaponry.[29]

The political-military relations during the then-routine times, that is, the general political situation at the beginning of the 1990s, should be distinguished from these relations at the time fighting broke out with Chechnya in 1994.

Considering the general political situation, the rising power of the conservatives in the government and in the Russian Parliament at the time indicated long-term army influence on the definition of the new national military policy of Russia. The first official military doctrine that was approved by President Yeltsin and the Russian Security Council in November 1993, clearly reflects the existing political-military consensus regarding military policy and its requirements. This consensus reflects the continuing emphasis on military power as the center of Russian policy in the international arena. Civilian and military leaders were in agreement that the military technological potential to reach parity and even superiority in the revolution in military affairs was the principal guarantor for Russia to maintain its status as a world power.[30]

However, in the mirror of the war in Chechnya, political military relations were characterized in both military campaigns by ups and downs, based for the most part on lack of consensus, fraud and deceit on the part of the political echelon in dictating reality in the field in order to force the opponents of this policy in the army command to carry out its orders. The senior army command expressed its desire to assist in solving the conflict, but only by using democratic means.[31] Against the backdrop of severe differences of opinion within the political establishment itself regarding a solution to the problem, the army urged politicians to settle the disputes among themselves and to accept a defined and clear position regarding what was expected from the military.

The incursion in December 1994 was not legal in the opinion of the army command: in opposition to the Russian constitution, the incursion took place without a declaration of emergency by the political echelon and with no consultation between the President and Parliament. In addition, the incursion took place before the expiration of the ultimatum that had been given to the Chechens by Yeltsin. There was no consensus, as well, regarding the readiness of the army. The senior army command felt that the army was not prepared to fight in Chechnya.

Fraud on the part of the political echelon in initiating the incursion continued with deceit in May 1996, as well, in the negotiations between Yeltsin and the Chechens. The motivation was election mischief. However, with his reelection, the fighting for Grozny was continued.

Leading up to the First Campaign

The political leadership was not aware of the sad state of logistics within the Russian forces; there was a rift between the political and military echelons based on unrealistic messages from Yeltsin, and fraud in hiding the Russian defeat by the political leadership. The lack of consensus within the political leadership itself regarding a solution of the dispute and a cease-fire deteriorated into a lack of consensus among the senior army command. When Yeltsin actually granted the army authority for a cease-fire, a rift appeared among the senior army officers that created an atmosphere of rebellion. Despite the orders of the political echelon for a cease-fire, bombing continued for a time.

During the Second Campaign

For Grozny, lessons had been learned not only in warfare tactics on all levels, but also in regulating relations between the political and the military echelons. It may be assumed that the increase in the intensity of the dispute with Chechnya, one of the first reasons of which was an army decision in the field to solve the conflict, created a more stable consensus between the political and the military echelons in a reality that they themselves had created. From 1996, the senior army command fell into line with the political echelon and adopted its consensus on the objectives and the continuation of the army fighting in Chechnya.[32]

Summary

The deceit of the political leadership is at the basis of the absence of consensus between the two echelons. Although the lack of consensus at the beginning of the conflict was primarily about the means to manage the conflict between the two states, while the senior command did not see an immediate means of solving the conflict with military action, the deception of the political echelon accords this variable unique power. The character of the developing warfare contributed the reduction of disagreement among the echelons.

The Effects of Relations between the Political and Military Echelons on Military Resilience

In the First Campaign—It has been found that damage was done to *Unit Resilience (Cluster B) in the ability to maintain the organic framework (B1),* in *maintaining the identification and the internal cohesion (B2),* in *coping with the burden of activity that leads to fatigue (B3)* and in *the organization of unit belonging and identity(B4).* The damage was done on the level of senior command and on the level of the simple soldier in the field.

On the level of senior command, there were serious disagreements. There were statements of enthusiasm and boastfulness that were expressed for example, in officers' utterances such as, "We need another Stalin to show us how to deal with those dark skins," "We don't have to just defeat Chechnya. We must destroy it," and "We could take Grozny with a brigade of parachutists in two hours." But alongside these, there were expressions of opposition within the army from other officers. Eleven generals from the Council of Land Forces protested against the action, maintaining that, considering the poor state of the army since the break-up of the Soviet Union, it was technically unready to undertake the mission.

The disagreements caused damage to the resilience of the units in terms of the disunity of the command that penetrated down to the level of the simple soldier. This was compounded by the fact that units, who were not fighting together, were transferred in haste from diverse areas and were placed under the command of senior officers who were responsible for their mobilization. However, these officers refused to aid in the mobilization of unit and submitted their resignations.[33] This refusal did damage to *Building of Force* (Cluster A) in *decrease in the extent of fitness, training, and the quantity of units in training (A1), damage to the framework of the training (A4); Military Framework (Cluster C)* in *the ability to carry out the orders and procedures of a professional army (C1),* and in *maintaining channels of internal and external organization (C2).*

In the absence of agreement about the military goal that was imposed from above by the political echelon-suitable preparation for the operation was prevented. Among soldiers who were sent to the campaign, there were new recruits and soldiers taken from armored units who had no experience in fighting on foot in an urban area. Most importantly, the lack of willingness among a group of generals to cooperate, prevented drawing conclusions about Russian army functioning since its participation in the war in Afghanistan. These generals were well aware of the failures in that war, but there was no correction of the problems. For example, when the Russians tried to conduct the battle in Grozny, there were structural failures in training and activation of forces as had occurred in Afghanistan. The Russian military doctrine involved activating armored infantry next to motor vehicles, usually at a radius of up to 200 meters. Russian infantry equipment, which included clothing and defensive flak jackets were in keeping with this doctrine, and thus, was quite heavy. It had already become clear in Afghanistan that the weight of the equipment caused the soldiers to move very sluggishly and, in effect, to prevent them from moving too far from a motor vehicle in order to carry out independent movement or to react quickly to threats. The damage to training imposed by the generals actually did not allow for the drawing of conclusions. In the campaign for Grozny, as well, equipment limited the Russian soldiers and made their fighting more difficult.[34]

The damage to carrying out orders and procedures was again on two levels. The senior command who opposed the incursion into Grozny refused to advance, even after they had received an order, and announced that "we will not use tanks against citizens . . . it is not our fault that we are here." Generals refused to accept command in the Chechen operation, even at the price of their dismissal. On the other hand, complete units, at the initiative of simple soldiers, refused to fight. A disruption of the channels of communication on all levels was the inevitable result of the split in the army, the severity of which increased as a result of the warfare strategy at that stage.

Damage was also caused to *Significance in the Realm of Military Spirit (Cluster D) in expressions of fear and anxiety, exhaustion, etc. (D1) and in questions about the significance of the mission (D2).* The atmosphere of disagreement contributed to demoralization: There were reports of the depressed emotional state of soldiers and nervous breakdowns, desertions, and even aid to the Chechens in destroying the Russians' own vehicles.[35]

The Second Campaign—As mentioned, the second campaign for Grozny was characterized by consensus between the political and the military leadership regarding both objectives and methods, which mitigated the intensity of the damage to resilience.

Summary of the Effects of Relations between the Political and the Military Echelons on Military Resilience

There was a lack of consensus between the two echelons, at first in the very objectives of the conflict with the Chechen nation and afterwards, this lack of consensus focused on the methods of managing the conflict. This led to great damage to military resilience.

Russia-Chechnya: Summary of the Connection between Military Resilience and Military Model, Warfare Strategy and Military-Political Echelons Relations

Relations between the military and the political echelons and relations within the army command itself constitute a variable with great effect on military resilience. The involvement of these relations in the variable of warfare strategy have a cumulative linking effect on the weakening of army resilience. The military model represents only a small amount of damage to resilience. These tendencies change from the first Russian campaign, which began in 1994, to the second campaign, beginning in 1999.

As stated, the relations between the political and the military are decisive regarding the resilience of the Russian army in the campaign on Chechnya. The refusal of the army to carry out the political decision to invade Chechnya, comes close to being a clear illegal refusal to carry out orders, as the decision itself was seen by the military command as an illegal order. The intensity of damage stemming from lack of consensus on the military was expressed on every level of command and on the actual opening of the campaign. In the second campaign, even though the Russians were victorious, there were still signs of damage to resilience. The cruel and barbaric acts taken during the fighting cast doubt on the effectiveness of implementation, and indicated a different aspect of mental dissonance in the warfare, and expressing damage to resilience.

The warfare strategy in the first campaign, which was a failure, contributed to the damage to resilience, as well. The attempt to correct logistic and tactical failures as they occurred in a kind of "patchwork" of ad-hoc acts did not allow for effective implementation and exhausted the fighters mentally.

However, when a warfare strategy was developed in which the army could direct the fighting to its own advantage the sense of confidence and operational achievement increased. As stated, the implementation effectiveness remains in doubt in light of the immoral dimension of the warfare.

Regarding the military model: The fact that in most of its variables, the traditional linkage between the army and the political echelons still ruled ensured that attention to the problems of resilience would mean decisions that were aggressively imposed, and not necessarily with liberal rationalization and possible discourse between the two echelons. Proof of the connection between the military model and its resilience is problematic in light of the crisis that the army was going through in the process of deep change on the political- social—and economic levels.

It may be asked how it is possible to distinguish between damage to military resilience, which was a result of the dynamics of the three areas that have been cited above and damage to resilience as a result of the variables described, that is, from the relations between society and the military, relations between the media and the army, warfare strategy and relations between the political and the military echelons.

An analysis of the professional literature demonstrates this clearly and enables specific identification of the damage and its link to the actual variable in warfare management in Chechnya.[36] Therefore, there should be differentiation between the weakness of the army in a period of critical change in Russia and the direct damage to resilience due to the warfare factors in the field during a period of operational activity.

It would be difficult to claim that the political leadership was concerned about the military resilience at the beginning of the conflict. Yeltsin's acts of deceit at the beginning of the first campaign and the sanctions and punishments used against the army command who opposed the incursion do not indicate worry about resilience. In addition, the absence of an attempt at a public explanatory campaign and the disregard for the role of the media in the conflict and its influence on both the society and the military testify that the political echelon took the support of both the public and the media for granted.

The change in the approach of the political leadership took place with the second campaign when steps were taken that show the concern of the politicians for public and media support, for the army and for world public opinion including:

- Granting first priority status to the media war in which the political leadership succeeded thanks to an efficient explanatory campaign.
- Tight control on all of the media.

- Establishment of the Russian Information Center, which controlled and filtered reports from the battlefield before they were distributed to the media in Russia from media and information sources abroad.
- Publicizing explanatory material which presented the Chechens as Islamic extremists using terror, a serious threat to the security of Russian citizens and which justified the war objectives, portrayed as inevitable.

Table 5.2 Russia-Chechnya 1994–2000

Military Model, Political-military relations, Warfare strategy	*Description*	*Military Resilience*
Military Model	Shift from late-modern to post-modern military	Low Damage
Warfare strategy	1994–1996 Integrated warfare (small and large scale)	High damage
	1999–2000 Large scale warfare	Resilience recovery
Politico-military relations	1994–1996 Non-consensus on aims and means	High damage
	1999–2000 Consensus on aims and means	Resilience recovery

Chapter 6

Israel-The Palestinian Authority 2000–2004

The Military Model—Integrated model: Modern, late modern, and post-modern[1]

Perception of Threat

Israel views itself as being exposed to three types of threat: enemy invasion, nuclear threat, and a sub-national terrorist threat. The intensity of each threat changes through time: The danger of invasion of Israeli territory has decreased in recent years, but the threat of terror has increased as has the potential for nuclear war.[2]

Force Structure—A large army based primarily on compulsory military service

The Israel Defense Forces is a *mass army* based on compulsory service and on the reserves, supported by a permanent professional staff.[3] Proposals to consider a voluntary armed force or some type of military service based on selectivity have met with opposition. Maintaining "an army of the people" is part of the national ethos.

Main Mission—Defense of the country

However, the violent dispute with the Palestinians has made it necessary for the army to fulfill an array of other tasks, such as humanitarian missions that are integrated with active combat.

Dominant Military Profession—From combat officer to soldier statesman

The dominant profession in the IDF is still the combat officer, who receives most attention. However, the development of a system of academic training programs, such as the "atuda," "talpiot," "atidim" and "mokdei yeda," testifies to the planned encouragement of military personnel who are scientists and educated soldiers. This stems from an awareness of the complexity of the combat field. Moreover, the dispute with the Palestinians has increased the need for a soldier statesman, who has been termed a "strategic corporal."[4]

Relations between Society and the Military—From support to ambiguity. Society is not indifferent, but is both supportive and critical.

It cannot be said that Israeli society is indifferent to the army. Citizens, or at least the families of those who actively serve in the army are involved in army life and thus, the army is not disconnected from the general population. However, practically, cracks and breaches have appeared in these relations. Some of those who should be candidates for compulsory service are not drafted, and among those who are drafted, no small number do not complete their full terms of service. In the reserve forces, this is even more striking due to the fact that that only a small number of reserve soldiers serve in significant ways. This breach has been deepened by the passage of the Tal Law exempting ultra-orthodox youth from compulsory military service.[5]

Admiration for the IDF, bordering upon enthusiasm and even adoration, and the fact that everyone serves as regular soldiers and then, in the reserves, have been well-known features of Israeli national identity. However, as time has passed, relations between the Israeli public and the army have almost reached a crisis point. Although the army is still at the head of national bodies that gain public trust, the status of the IDF does not appear as secure as it was in the past. Criticism of army activity, which was once the sole prerogative of a small circle of experts, has expanded into the public sector. It seems that civilian courts, the media, and parents of soldiers now compete in expressing criticism of army behavior and administration.[6]

During the period of the present research, the combat methods used by the IDF during the conflict: use of snipers to strike at armed opponents, use of tanks and APCs in urban areas, tactical use of bulldozers to gain an advantage in the combat arena, and air support by unmanned aerial vehicles and air force combat helicopters- were criticized due to the great destruction left behind, and to the harm caused to innocent civilians. Some have claimed that the IDF was using "exposure" operations (a term used by the IDF for an operation in which an area is completely cleared of obstacles and any objects that might block the field of vision or might serve as cover to terrorists—including vegetation, various structures and orchards) as a means of punishment in order to pressure the Palestinian population to stop cooperating with those who fired Kassam missiles and terrorists, as psychological warfare.

Israeli society is split into two camps regarding the policy of targeted assassinations carried out by the IDF: Leftist groups feel that even considering the violent terrorism Israel was subjected to during the Al-Aqsa Intifada, as a law-abiding state, Israel was required to put terrorists to trial and to let the courts determine their punishment. The decision to put someone to death should not be left in the hands of a limited number of ministers and Defense Ministry staff. Even if those assassinated were murderers, they should only be killed in a situation of clear and immediate self-defense. Assassination without trial is murder, even if the victim is him/herself a murderer. Moreover, assassinations are not carried out only against those who are defined as "ticking bombs" but also against members of the Palestinian resistance who are politically active and not directly involved in terrorism. At times, the assassinations are meant to serve as deterrence or revenge, or for political purposes. For the left, terror organizations react to targeted assassinations with the most violent acts of revenge and thus, avoiding assassinations and concentrating only on acts of defense will save lives in Israel.

On the other hand, many in Israel, support targeted assassinations for the following reasons:

- Anyone who enters the bloody cycle of terrorism and who murders innocent civilians cannot expect to have the protection of the law as other citizens do. Leaders and activists in terror organizations cannot be allowed to continue to organize attacks in the name of civil rights and humanistic perceptions.
- Jewish morality sides with the principle that "if someone comes to kill you, kill him first," especially when referring to a state of war and to murderous acts of terror. Sometimes it is impossible to trap terrorists, or to gather admissible evidence in a court of law regarding the use of terror because members of terror organizations are hidden in hostile and problematic areas where the police cannot arrest them.
- In practical terms, terror is threatened only by counter-terror. This was the case when the British acted against pirates or when the United States operated against Al-Qaida. Those who bow their heads before terror are hit harder, as the objective of terror is to induce fear among the civilian population. Thus, the harder terror is hit, the more people on the opposing

side will fear to join the cycle of terror when they see the bitter fate of their leaders.

In any case, regarding military operations (which will be analyzed in detail in the discussion of the combat strategy variable), the decision of the Israeli government to initiate the "Defensive Shield" operation received almost complete support among the Jewish population of Israel. This wide agreement was reflected in the high response rate to the call up of the military reserves and in the high motivation among the mobilized soldiers.

The ultimately sharp criticism of the operation centered around army activity in the battle for Jenin, which led to a great number of casualties among Israeli soldiers (23 killed, 75 wounded). A number of the bereaved families called upon the then-Israeli Minister of Defense Benjamin Ben Eliezer to resign and accused him of abandoning their sons after he admitted that he had refused to approve an attack upon the camp from the air, fearing for the lives of Palestinian civilians.

Among the Israeli public, "Defensive Shield" contributed to raising morale after a difficult period of terror attacks. All of the reserve soldiers who had been called up and those who volunteered for the operation reported for duty and many Israelis felt that the operation was a "fight for their homes." On the left, there were some who opposed the operation and criticized the great suffering that it caused to Palestinians. On the other hand, those on the right claimed that the operation proved that it was possible to stop terror using military power.

"Operation Rainbow" was also sharply criticized by leftist organization after the "Palestinian demonstration affair" (On the second day of the operation [May 17th], 10 Palestinians were killed while demonstrating, as they marched towards Tel-Sultan. They were apparently the victims of a tank shell that had been fired at an empty building but was not stopped by the target and continued on). There were demonstrations in Tel Aviv against the Ministry of Defense and activists from the peace organizations, the Peace Bloc, ISM and "The Courage to Refuse" clashed with police and destroyed army equipment near the Kissufim crossing point. The B'tselem human rights organization accused the IDF of a war crime. The army expressed regret for the incident and maintained that the demonstrators had been responsible as they had purposely entered a fire zone. In addition, the army suggested that the shell that had been fired- not at the demonstrators- might have caused a chain reaction of explosives that had been placed there by Palestinian terrorists, and that is what had caused the great damage.

These changes in relations between society and the military represent the army as being based on modern and late modern military models, but by no means a post-modern model. During the period of this research, relations ranged between strong support for the army and criticism of its actions.

Army–Media Relations—From being manipulated to being sought after

The Israeli media sometimes feels manipulated, but most striking is the fact that the IDF is aware of the role of the media and its importance. This sense is sharpened in limited conflicts during which the IDF seeks out the media for explanatory purposes.[7]

At the beginning of the Second Intifada, the media also received information from other sources: Senior officials of the General Security Services (Shabak), senior officials of the police department, American sources, Palestinians, Arabs and others. The army transmitted a policy of restraint and the media repeated this message. At this stage, the media broadcast messages that came from meetings with the Prime Minister, the Minister of Defense, or with the Chief of General Staff, and presented these as fact. The media served as a propaganda vehicle for the government. There were only two exceptions: From time to time the media transmitted a message of anger with the government that seemingly came from the public—and this message was almost always more right-wing than the official position; and from time to time, once or twice during the Intifada, when the American government expressed a strong position in opposition to Israeli government actions, the Israeli media adopted the American position. This is what happened in June 2003at the start of the "road map" dynamic.[8]

During the period of the current research, this tendency continued, along with expressions of severe criticism regarding the military management of the conflict. It was impossible for the media to ignore the reports frequently issued by the IDF, especially during the various operations.

In the battle of Jenin, during Defensive Shield, the media was inflamed when the IDF was accused of massacre. The army reported that it had found only 46 bodies, of which five were booby-trapped, according to army spokesmen. The great majority of victims had been armed. But the media criticized the exaggerated use of force, especially regarding the widespread use of bulldozers and the complete destruction of the Hawashin neighborhood. In addition, they maintained that human rights had been thoroughly violated, medical aid had been prevented from reaching the victims in the camp, mass arrests had been carried out and much civilian property had been destroyed. The army responded that it had done everything possible to avoid harming civilians, including the decision not to bomb from the air, and that the great destruction carried out by the bulldozers had been necessary. But in an interview with one of the bulldozer drivers published in Yedioth Aharonoth, a mass circulation daily newspaper, the headline read "I created a stadium for them in the middle of the camp"[9] that put this claim in doubt. On the other hand, in a news report headlined "Into Dust," which had been published two weeks earlier,[10] the bulldozer drivers asserted that they had acted with the greatest restraint and explained that if they had been uncontrolled, they could have done more damage. In addition, harsh criticism was directed towards the IDF spokesman and the decision of the army to close the refugee camp in Jenin to news reporters for two weeks.

Later the army succeeded in mobilizing the media in order to counter Palestinian claims of massacre in Jenin. The Palestinian version of what had happened in Jenin was revealed as false after investigations by the United Nations and by human rights organizations that rejected the accusation of massacre. Today, there is wide acceptance of their findings on the part of the international community[11] and the claim of massacre is limited to Palestinians, to Arabs who get their information from the Arab-Muslim, anti-Israeli media and left-wing pro-Palestinian groups. An IDF unmanned drone documented a funeral in Jenin after which the "deceased" stood up and left. The IDF maintains that these photographs prove that the Palestinians tried to fabricate evidence in order to support their fraudulent claims.

Conscientious Objection—Prohibited by Law

Against the backdrop of the crisis involving the concept of "a nation in uniform" as it affects the perception of compulsory service—the transformation of the IDF from a mass army to a small professional military contradicts the basic assumptions of the mainstream in Israeli society, according to which compulsory service is connected to the very survival of the Israeli nation-state. Thus, conscientious objection is not an option. Although the motivation to serve in the IDF, especially if measured by the readiness of 18-year-old conscripts to serve in combat units, is still much higher than usual in armies throughout the world, there is an increasing change in attitude and behavior patterns towards service.

There have been claims that the rise in conscientious objection during the period of this research and the gradual change in public attitude towards refusal to serve have been affected by the growing politization of the army. At present, conscientious objection is prohibited by law and there is no alternative civil service for conscientious objectors.[12]

A summary of the model army typology: When considering threats, the IDF fits the three types of armies. Regarding force building, it belongs to the traditional types, but selective mobilization, especially for the reserve service, testifies to a different trend. In other words, the IDF differs in behave and in construction. Most of its quantitative strength is based on the reserves, and to that extent, it is a modern army. However, the selectivity in mobilization, that is, the non-utilization of all of its potential power testifies to a move away from a modern army and towards a late modern and post-modern military model.

The Effect of Military Model on Military Resilience

A wide range of professional literature deals with this dilemma against the backdrop of the question of transforming the IDF to a post-modern army.[13] Methodologically, it does not appear that details of the weak links of the IDF discussed in this literature can all be included in the framework of harm to resilience during the period of conflict discussed in the research. This is true for two reasons:

- There is no consensus in the professional literature regarding the characteristics of IDF weakness in the existing model.
- Even if these weak points realistically do exist, their development is not specific to the period of conflict but rather is expressed as a reflection of all of Israel's recent wars.

Thus, it is important to focus on harm about which there is consensus and which is relevant to the management of the conflict under discussion. However, there is no factual data for this period due to military classification of information. This relates to all the components of *Force Building* (*Cluster A*).

This damage is caused by two factors that are different but that have mutual relations: the crisis in motivation and the exhaustion of soldiers serving in the regular army and those who are candidates for conscription on the one hand, and greater utilization of reserve forces on the other.[14] The erosion of motivation is clear and has been identified primarily in the Westernized secular middle class population, including kibbutz youth. This is expressed in a decrease in the readiness to be drafted in general, and specifically a decrease in the willingness to be drafted into combat units, and therefore, the reduction in selectivity for combat positions, a decrease in the willingness to volunteer for officer's courses, a rise in the number of those lowering their medical profiles, and a rise in the number of those requesting to serve near home.[15]

Regarding the reserves—Despite the high motivation that characterized these forces during the period of the research, the motivation was not firmly based. The burden of reserve service increased due to the need to protect settlements in the occupied territories. The lack of monetary recompense led to new expressions of "motivational crisis" on the part of reserve soldiers and to a wave of new protests that came close to becoming a revolt. This protest was politically exploited by the Shinui Party.

The damage to *Force building* can be understood against the backdrop of the importance of the reserves, as, in Israel, a large part of the available human resources mobilized in times of war are reserve soldiers. The unique qualities of these forces are especially relevant to components of force building such as, for example, their large numbers, their low cost, their moderation, and the fact that they provide a balance between the need (which may be real or apparent) for a massive army and the desire to maintain a state with normal life.[16]

Fishman[17] points to the decreasing numbers of reserve soldiers, the fact that the reserves barely train, are not technologically up to date and are not ready to deal with existing threats. Fishman states that in the last five years, the numbers of reserve soldiers have fallen by 20 percent and that more than 30 percent of reserve soldiers have not trained at all in the past four or five years. Not only is this force, which by definition represents the main mass of land forces, getting smaller but it is also becoming less qualified and less well-suited to deal with current threats. According to Fishman, there is a sense that the reserve forces are gradually atrophying.

The fact that cardinal security decisions are made solely on the basis of their results, due to budgetary distress resulting from the changes in the attitude of society to army service in general and to reserve service in particular, casts a shadow over the links between threat scenarios, the number of potential fronts and the required training for conventional combat, while at the same time managing a low intensity conflict. Fishman stresses that even today the standing army has "run ahead" of the reserves and increasing gaps have opened in professional training and combat methods. These gaps begin in the realm of precise fire ability, include the use of advanced intelligence methods such as unpiloted drones with which reserve units are less acquainted, and finally, methods of command in which the technological gaps are particularly striking.

Fishman feels that, during the Intifada, combat theory regarding the struggle against terrorism that created different training systems for the regular army and for the reserves, led to the development of special means of combat for the regular army, while the reserve system continued to train for conventional warfare. In essence, two armies were created. One of these was based on conventional combat to be used agaisnt neighboring states, a threat that was perceived by the command as remote, while the second was based on the battle against the threat of terror and against more distant states.

Fishman maintains that when the security establishment is asked: "What are the red lines limiting the decrease in the size of the force, and what are the alternatives to a reduction in its size and in the operational readiness of the reserve system, the answer is: technology. In Fishman's opinion, this is "the new deity that everone worships."[18]

Regarding the effects of the media-army relationship on military resilience, as mentioned, there is no data presented that might be used to conclude that any damage has been caused. However, the fact that a severe image problem is created when the information from the field is provided by military men in uniform should not be ignored. The very fact that informing the public is a matter of expertise that has not been included in any aspect of military training and that information is presented by a person in uniform creates a barrier between the army presenter and the media. The inflexible structure of the army unit makes the activity of the unit entrusted with information transmission much more difficult than it would be for a civilian/governmental unit responsible for transmitting information.

Summary of the Effect of Military Model on Military Resilience

The Israeli military model is composed of characteristics of three models that damage military resilience to a minor extent.

Warfare Strategy in the Conflict

The principal feature of the Al-Aqsa Intifada was the wave of terrorist attacks carried out by all of the Palestinian terrorist organization and even by women and children. Up until the end of 2001 the Israeli government adopted a policy of restraint with Prime Minister Sharon asserting that "restraint is strength." But this restraint did not lead to calm. The lack of action on the part of the Palestinian Authority and the additional acts of terror during the summer (the most serious of which took place at the Sbarro Restaurant on August 9 in Jerusalem when 15 people were killed) led very quickly to a loss of faith in the Palestinian Authority and a demand for strong military action against terrorism.

The government decided on a policy of supervised escalation and began to carry out short incursions into Area A, the area under Palestinian Authority control. In each incursion a number of wanted people were arrested and a number of structures belonging either to the Palestinian Authority or to terrorist organization were destroyed. Additionally, a number of wanted senior members of Hamas, Fatah and the Popular Front were assassinated.

In 2002, terrorism again increased and reached a peak in "black March" during which more than 135 Israelis were killed in suicide attacks that occurred daily. The most serious attack was the massacre at the Passover Seder in the Park Hotel in Netanya when 30 people were killed: Most of them were elderly and several were Holocaust survivors. After this attack, the Israeli government initiated the Defensive Shield Operation against terrorism.

The examination of warfare strategy starts from this point and focuses on the operational level and military actions that have been carried out since then in order to examine the variable of combat strategy.

From March 2002 to May 2002—The Defensive Shield Operation was a wide-ranging campaign of the IDF that began on March 29, 2002 in Judea and Samaria, and was officially ended on May 10, 2002. The principal objectives were to harm the Palestinian "terrorist infrastructure" and to stop the wave of intensifying attacks. In the operation, the largest number of reserve forces were called up since the end of the Peace in the Galilee Operation.

The Geographic Conflict Characterized by Blurring of Boundaries, Unstable Contacts, but Close Combat

The geographic area of conflict included the cities of Ramallah, Shechem, Jenin, and Bethlehem, that is, in built up urban high population density areas. In Ramallah, there was hand-to-hand combat, but the IDF did not meet with great resistance and an elite unit of paratroopers laid siege to the Mukataa, containing government buildings and the headquarters of Yasser Arafat. In Shechem, there was variable combat with Palestinian resistance and hand to hand battles.

The most severe fighting took place in Jenin. The Palestinians booby-trapped the camp and prepared ambushes in the narrow streets of the city and the IDF forces conducted house to house fighting and hand to hand battles.

The large number of terrorists who had assembled at the center of the Jenin Refugee Camp resulted in many casualties among the Palestinians and to great destruction in the camp after it had been decided to destroy every house that was suspected of harboring terrorists. The procedure for destroying houses included a warning announcement to evacuate the house and to enable suspects to give themselves up before the bulldozer began to dismantle the house. In Bethlehem IDF forces had to deal with suspects who had sought refuge in the Church of the Nativity.

Man-machine Interaction Characterized by the Human Factor

The ambush method carried out by the terrorists, the camps rigged with explosives, the snipers, the booby-trapped houses and the massive use of armored bulldozers by the IDF characterized the human factor in Operation Defensive Shield.

Professionalism versus Popular Involvement—Participation of Civilians

The IDF was fighting Tanzim forces, the Islamic Jihad, al-Aqsa forces and the Fatah, and there was also great involvement of civilians whose houses had been booby-trapped and were subsequently destroyed by IDF bulldozers. The IDF reserves, reinforcement forces from the regular Army Battalion 51 of the Golani Division, tank units, Navy Unit 13, the Yahalom commando unit, with sporadic assistance by military helicopters, armored vehicles and D-8 bulldozers all provided the professional Israeli dimension, fighting against popular combat of Palestinians, which included primarily improvised land mines, booby-trapped houses, sniper fire and RPG rockets.

Combat Method—Asymmetry between the sides in means of combat management combined with tactical maneuvering and minimal fire, sometimes without movement

The principal means of combat among the Palestinians preceding Operation Defensive Shield had been explosive belts provided to suicide terrorists who were to blow themselves up in crowded public places at the centers of Israeli cities. In contrast to the preceding Intifada, the strict security measures imposed by Israel led to the use of women and children as suicide terrorists. This resulted in harsh criticism both internationally and in Palestinian society, but the criticism was not strong enough to end the bombings. The suicide attacks were the main reasons for the death of Israeli victims.

In addition, Palestinian terrorists were armed with light weapons that were used to attack soldiers and settlers in the occupied territories while traveling by car. The following is a list of combat means that were widespread among Palestinians:

- AK47 Kalashnikov rifles, most of which had been smuggled in.
- M16 rifles that had been stolen from Israel.
- Machine guns, sub-machine guns, hunting rifles and sniper rifles of different types.
- Molotov cocktails (used primarily by children and youths).
- Pipe bombs (used primarily by children).
- Hand grenades.
- RPG-7 rocket launchers with homemade bombs and possibly with RPG standard explosives.
- Al-Batar and al-Bana antitank missiles (usually home-made).

The Palestinians also often made use of explosive devices: roadside charges, improvised land mines and EFPs against IDF armored vehicles.

During Operation Defensive Shield, the tactical maneuvers that were used by the IDF were the result of the urban combat conditions and their objectives were the removal explosives, the destruction of booby-trapped houses and the surrender of Palestinian fighters; the bulldozers were dominant during these missions. The central movement of IDF forces focused on the road between the four cities, but was limited to attacking specific strongholds until these surrendered.

Warfare Strategy after Operation Defensive Shield

During the years of conflict, the IDF developed and perfected combat tactics for fighting in urban areas including the use of intelligence, the use of snipers against armed Palestinians, the use of tanks and other armored vehicles in urban areas, operating bulldozers tactically to gain an advantage on the battlefield as well as air assistance by air force drones and military helicopters. These combat methods were successful in lowering the number of attacks against Israelis and reduced casualties among IDF soldiers, but were sharply criticized due to the great destruction that left behind and the harm to unarmed civilians.

Another IDF method that became common during this conflict was the targeted killing of Palestinian operatives. Israeli intelligence agencies identify leaders and commanders of terrorist organizations and according to the information received, forces are sent out (usually air force helicopters) to kill them. The official Israeli claim is that the "targeted killing" prevents the murder of innocent civilians. In actual fact, many civilians are killed during these operations. This method has created demoralization among Palestinian organizations and has disrupted their activities, and this, apparently, is its military benefit.

Regarding activities in the territories—including arrests of suicide terrorists before they have attacked and the targeted killings of terrorist leaders—a strong link between the IDF and the General Security Services has developed in which the GSS supplies intelligence information to the forces in the field. Border guards and the police force also take part in anti-terror activities in the territories.

Warfare strategy was different in Operation Defensive Shield, which was a more intensive military operation, but the Days of Penitence Operation in Sept. 2004 was more characteristic of normal military combat with fewer features of small-scale combat.

Operation Rainbow—May 18th through May 25th, 2004

Operation Rainbow was a military campaign conducted by the IDF in the Rafiah area and along the Philadelphia Corridor from May 18th, 2004 to May 25th, 2004. The campaign began a few days after an M-113 Engineering APC was blown up by Palestinian anti-tank fire and its five-man team was killed. The objective of the operation was to act against the terror infrastructure of Rafiah, to kill as many terrorists as possible and to create a safer environment for soldiers patrolling the Philadelphia Corridor. An additional goal was to expose and to destroy smuggling tunnels and to prevent the delivery of Sager anti-tank missiles and Sterla shoulder-fired anti-aircraft missiles that were awaiting smugglers in the Sinai Desert.

The geographical conflict was still characterized by fluid boundaries, but the IDF isolated Rafiah from the rest of the Gaza Strip and set up large earthen barriers, and soldiers were stationed around them. This was meant to establish boundaries between the sides, although fluid contact and hand to hand close combat were still dominant, such as the fighting that took place in the Tel-Sultan neighborhood. *The rise in the use of the machine factor in the fighting occurred* when the IDF was reinforced by a battalion from the Golani Division in Achzarit APCs and Merkhava tanks, and the Palestinians brought anti-tank rockets and missiles into operation. The *intensity of fire increased.* There was also *great involvement of civilians* in the "demonstration" incident."

The Days of Penitence Operation- September 30th to October 15th, 2004 was conducted by the IDF in the Northern Gaza Strip between Sept. 30th and Oct. 15th, 2004. The operation focused on the areas of Beit Hanun, Beit Lahia, and the Jebalia Refugee Camp, which were the principal sites from which Kassam rockets were launched to attack the city of Sderot and the Israeli settlements in the area. The operation began as a reaction to the deaths of two children in Sderot during a Kassam rocket attack. Five Israelis were killed during the operation (three civilians and two soldiers) and between 104 and 133 Palestinian died, most of them terrorists but between 18 and 31 children. (There is no agreement on the data.) Approximately ten Kassam missile teams were destroyed (each composed of 2 to 5 terrorists and 1 to 5 rockets). In addition, 77 houses were destroyed and a large number were damaged during the fighting. The Israeli Engineering Corps exposed wide areas that had been used as hiding places for rocket firing and demolished houses from which anti-tank missiles, mortar bombs, and Kassams had been fired.

In this operation the IDF used the adjacent and precise air support of remote controlled drones and combat helicopters that identified and destroyed concentrations of Kassam rockets and their teams and anti-tank missiles, while ground forces pinpointed targets, searched areas and tempted terrorists to initiate contact. During the operation, dozens of anti-tank rockets were fired at the IDF but they caused only light injuries, which testified to the high quality of protection afforded by the Israeli armored vehicles.

The warfare strategy during this operation was still characterized by *permeable borders* due to the fact that fighting went on in the built up urban areas of Beit Hanoun, Beit Lahia, and the Jebalia Refugee Camp. However, there was a *decrease in variable contacts and in face to face combat* due to the aforementioned air assistance by remote controlled drones and military helicopters, which identified and destroyed Lassa, and anti-tank teams while ground forces pinpointed targets, searched areas and tempted the terrorists to initiate contact.The Palestinians fired dozens of anti-tank rockets, mortars and Kassams.

The machine factor increased in the fighting:

- The IDF for the first time demonstrated the "bubble tactics" for fighting in built up areas. These tactics involved ground forces working with air

assistance that had the ability to make exact surgical strikes at terrorist targets marked from the ground, and at the same time, the drones and the air force helicopters relayed exact images of the battlefield to the ground forces. These tactics had great success and the Israeli Air Force was responsible for heavy losses of Palestinian terrorists and damage to Kassam lauchers.
- The use of protection devices on armored vehicles proved themselves effective against the threat of anti-tank weapons such as RPG-7s.

Combat was still characterized by professionalism juxtaposed with simplicity, although improvements in the combat against Palestinians could be identified; it was distinguished by minor involvement of civilians, the continuation of the asymmetry between the sides regarding the means of combat management (the Givati division, Golani battalions, tanks, engineering forces and the air force against groups of armed Hamas guerrillas, Al-Aqsa battalions, popular opposition committees and the Islamic Jihad), a large amount of fire power with movement, including airborne activity.

In a summary of the military operations of the Al-Aqsa Intifada, the dimension of combat professionalization was salient. The IDF developed special tactics for combat in built up areas. Some were meant to be used in the capture of an entire city while others were used for specific operations such as the arrest of terrorists. The IDF tactics were affected and improved by the army experience in Palestinian cities. These methods were based on a high quality of human resources, advanced technology and multi-unit coordination among all forces in the area.

These combat tactics of the IDF proved to be effective and this was expressed in the especially low number of casualties in battles with the Palestinians (about 250 soldiers were killed between 2000 and 2005, during the Al-Aqsa Intifada). The IDF first entered the high population density city centers and Palestinian refugee camps during the Defensive Shield Operation in 2002. Since then, the IDF has entered Palestinian urban areas thousands of times in both armored and other vehicles and on foot, to prevent terrorist attacks and to arrest suspects.

Some of these effective tactics have been:

- Obtaining good intelligence information on the operational area and the militias who operate within it. Obtaining intelligence by spying and by advanced surveillance as well as activating agents and informers who are under the responsibility of the General Security Services.

- The "swarming method": Non-linear progress in the operational field and attacking terrorists from every direction.
- Use of tanks and heavy APCs to safely transport ground forces to the center of the city while supplying protection against enemy fire and anti-tank rockets.
- Use of M109 and M163motorized anti-aircraft artillery in order to neutralize sharpshooters on the roofs of houses.
- Use of D-9 armored bulldozers to open traffic arteries for armored vehicles and for soldiers, to clear areas of explosives, to neutralize sudden traps, to break through obstacles and to destroy houses that were used to fire at Israeli troops.
- Making use of technical advantages such as night vision equipment, and advanced means of control and supervision.
- Use of ceramic protective vests that have proven effective in blocking bullets and in saving lives.
- Use of air support by remote piloted drones to provide intelligence and wide-ranging imaging of the combat arena.
- Air support of air force helicopters for look outs, for rescue operations, for air cover and to ensure precise rocket fire against RPG teams.

The most difficult battle of the operation was the one in Jenin in which 23 soldiers were killed. In other battles, such as those in Nablus, Kalkilia, Tulkarem, and in the Gaza Strip, IDF operations in Beit Hanoun, Jebalia, Khan Yunis and Rafiah (The Rainbow Operation), the IDF suffered few casualties while they killed hundreds of terrorists and arrested thousands.

Summary of Warfare Strategy

Small scale combat characteristics were prominent at the beginning of the conflict that continued, developing into professional operational combat including more characteristics of regular military combat.

The Effect of Warfare Strategy on Military Resilience

The fear of harm to civilians and the resulting limitations on the use of firepower that were salient especially at the beginning of Operation Defensive Shield led to damage to *Unit Resilience (Cluster A)* in *difficulty in dealing with overburdening activity led to exhaustion; Significance in the Realm of Military Spirit (Cluster D)* in *expressions fear and anxiety, exhaustion, among other effects (D1).*

The concern shown by the IDF for civilians in Jenin, and the small scale frugal use of combat force leads to a number of conclusions. In the Defensive Shield Operation, the army used armored forces, ground forces, attack helicopters that fired shells and rockets towards specific objectives, but the artillery remained silent. The commander of the battalion that was responsible for the sector refused to shell the refugee camp for fear of civilian casualties. "I could have finished it with a whistle," he said. "Divisional fire into the center of the camp, and the whole story would have been finished. But we behaved differently."[19] The combat planes remained on the runway as well. The Head of the Central Command, Yitzchak Eitan, refused the requests of those under his command to engage the air force in the battle "except for the controlled use of helicopters," saying that "there are innocent citizens here . . . if we use aggressiveness, many of them may be harmed."[20] Indeed, Israeli planes did not shell Jenin, or any other of the cities although such a move might have ended the battle. Moreover, even tanks were put into action only at a relatively late stage, despite the fact that the ground forces greatly needed their help. It appears that the commanders of the operation feared that use of heavy arms in the narrow alleys of the camp would cause great damage and they preferred to postpone the use of these armaments. The tanks that were put into action at the end of the battle were mainly used as armored bases for firing and "we not permitted to use artillery . . . for fear of uncontrolled damage."[21]

These limitations, that the IDF placed upon itself, certainly prevented serious injury to the civilian population of the refugee camp. On the other hand, they also put the soldiers' lives in danger and created a sense of clear anxiety and exhaustion. "We could have finished it a lot sooner," a reservist complained, "but we had strict orders not to throw a hand grenade into a house without making sure that there were no civilians inside first." Many soldiers felt resentment at the moralistic considerations that tied their hands and made them especially vulnerable. "We are behaving in Jenin like pizza delivery boys, who must come straight to the door of the terrorists' homes." said one of the Israeli soldiers. "We shouldn't get killed just because of the code of "purity of arms." The Palestinians, it become clear, actually tended to agree with this evaluation. One of the terrorists who participated in the fighting, Ta'abet Merdawi, said in an interview with CNN that he "and other Palestinian fighters had expected Israel to attack using planes and tanks." The surprise that awaited them made them very happy. "I couldn't believe my eyes when I saw the soldiers" he said. "The Israelis knew that every soldier who entered the camp would be killed.[22]

Here damage was caused to *the Military Framework (Cluster C)* in *limited ability to carry out rules and procedures of a professional army (C1); Corps Spirit (Cluster D) in questions arise as to the significance of the mission(D2); Realm of Operational Fitness (Cluster E) in accumulated burden of operating technical equipment, which created dilemmas about maintaining service norms and fitness(E1).*

Testimony of soldiers about their activity at roadblocks including acts of violence and humiliation of Palestinians and the admission of killing the 13-year-old girl, Iman al-Hams, the humiliation of the violinist Wissam Tayem, and the killing of Mahmoud Kamil testify to the damage caused. What these three cases have in common is that they aroused great public reaction in Israel. The admission of killing al-Hams, a 13-year-old girl from Rafiah horrified the public. The image of the violinist Wissam Tayem who played at the roadblock at Beit Iba, standing opposite the soldiers and playing his violin, and the soldiers firing at the wanted Islamic Jihad member Kamil at Kfar Rahbah near Jenin after he had been disarmed led the Chief of General Command, the commander and chief of the army, to tell a reporter from Ha'aretz newspaper that same day that "we must carefully investigate whether we aren't transmitting contradictory messages to the combat units."[23]

Additional evidence of the damage caused by the type of warfare in this conflict can be found at many Internet sites. Due to the untrustworthiness of some of these sites, we can indicate only what they have as a common denominator for soldiers' testification. Soldiers testify to looting, to destruction of property, to not obeying orders about when to open fire, abusive behavior and other similar acts.[24]

Damage to resilience has been expressed in the appearance of conscientious objection during service. It can be identified in a letter written by 27 air force pilots expressing opposition to carrying out orders that they considered immoral and to the demands of combat behavior that cast doubt on the norms of morality.[25] "The combatants' letter" written in the name of the "Courage to Refuse" movement (composed of 51 reserve officers) also attests to this harm.[26] In a letter from the elite Matkal (General Staff) Unit (13 reserve soldiers) the issue of harm to the army mission is stressed: ". . . We will not renounce our mission as soldiers of the Israeli Defense Force.[27] There is no numerical data regarding the scope of conscientious objection in practice. From the letters that have been mentioned above, we can estimate that the scope is limited. However, media resonance is especially powerful and thus, the damage to military resilience is great.

Summary of the Effects of Warfare Strategy on Military Resilience

Management of a conflict that included most of the characteristics of small scale combat led to great damage in military resilience. The transition to regular military combat moderated this damage although dilemmas in the realm of maintaining service norms and operational fitness continued.

Relations between Military-Political Echelons

There was consensus about both objectives and the means between military and political echelons, while, at times, there were limited shifts in consensus regarding means. The salient characteristic in these relations during the al-Aqsa Intifada: The military echelon recognized the authority of political echelon, so that when there was a difference of opinion between the two echelons, the military echelon obeyed the directives of the political echelon.

In this relations, the period should be broken down into the two terms of office of the two Chiefs of General Command: the term of Shaul Mofaz and that of Moshe Ya'alon. During the Mofaz term of office, there was full consensus between Mofaz and Prime Minister Ariel Sharon, who established a national unity government with Benjamin Ben Eliezer as the Minister of Defense. The personal composition of this government led to a unity that enabled new political positions to be decided upon for what seemed to be 'low intensity warfare." The three participants of the triumvirate considered that "there was no Palestinian partner," and that policies should be designed to "etch on the consciousness" of the Palestinians that they would not gain their objectives using terror and that only after this "etching" had taken place would it be possible to take political steps. Nevertheless, policies were not always carried out with complete agreement. On the 15th of October, 2001, Ben Eliezer expressed the opinion that Mofaz should leave his position after Mofaz had criticized the government decision to pull the IDF out of the Palestinian neighborhoods of Hebron. Despite the relatively sharp confrontation, this did not lead to a practical crisis and cooperation based on similar attitudes continued between the two. The new government began an aggressive policy that was principally based on "focused eliminations," IDF entry into all areas of the Palestinian authority, and heavy handed tactics against the insurrectionist Palestinian population.

On March 29th, 2002, the IDF initiated the very largest operation up to that point in the Palestinian uprising, the previously discussed "Operation Defensive Shield," which involved entering the Palestinian cities of Bethlehem, Ramallah, and Jenin and combating Palestinians who offered resistance. An important landmark in the fighting occurred with the deaths of reserve soldiers in an ambush in Jenin as they were battling in a dense urban area, and there were claims (which were later disproven) that a massacre had then taken place among the population of Jenin.

At the beginning of the operation, an open microphone recorded Mofaz whispering to Prime Minister Ariel Sharon that this was the opportunity to expel Arafat. Whether or not this was actually his position, it was not accepted and Arafat remained besieged in his office at the Mukata Compound in Ramallah.

During the term of Moshe Ya'alon as Chief of General Command, there was still consensus between the military and political echelons regarding the objectives of the conflict with the Palestinians and the military management of it, but there was less consensus about the methods of management. There were shifts in these relations, especially between Yaalon and the Minister of Defense Mofaz. Yaalon was chosen as Chief of General Command in July 2002, two months after the Defensive Shield Operation. His major task was to continue the war against Palestinian terrorism and to contain it.

In the first two years of his term, the wave of terror emanating from the West Bank ebbed, to a great extent thanks to the assistance of the General Security Services and the new security fence. The cities of the Sharon area and the north, which had suffered from many suicide attacks in previous years, enjoyed relative quiet throughout 2004. However, the Jerusalem sector remained almost completely open and there were a large number of severe attacks in the city.

In the Gaza Strip, difficulties escalated during 2004 and many Israeli settlements, including the city of Sderot, which is located within the Green Line, were under the constant threat of Kassam rockets which were launched from Gaza. Under Yaalon's direction, a number of large operations were conducted in the Gaza Strip (although even the largest of these, the Days of Penitence Operation was not similar in scope to Defensive Shield), during which IDF ground forces entered the Zaytun neighborhood, Rafiah, Beit

Hanoun, Jebalia and Khan Yunis in order to destroy Kassam building facilities and laboratories where explosives were made, and smuggling tunnels. In these operations, hundreds of terrorists were killed but dozens of Israeli soldiers died as well. In Rafiah, the IDF destroyed hundreds of houses in an attempt to damage the smuggling tunnels under them, but was only partially successful. Human rights organizations sharply criticized the destruction of Rafiah and rejected the IDF claim that it was a military necessity.

Simultaneously, Yaalon was forced to deal with difficulties that were not operational in nature. The dismissal of the commander of the Gaza division, Brigadier-General Shmuel Zakai, after he had failed a polygraph test regarding information leaks to the media during the Days of Penance Operation, aroused great dispute among the public and the political echelon. In addition, Yaalon had to deal with continuing criticism regarding the erosion of combat morality in the IDF during the Intifada as expressed, for example, by the "verification" of the killing of the young girl, Iman al-Hams and in the abuse of the corpse of one of the terrorists, in addition to the friction with the Palestinian population at the checkpoints.

Although the choice of Yaalon as Chief of General Command had been considered "non-political," Yaalon became involved in a series of embarrassing incidents in the context of various comments he had made that were entirely political in nature. Among these, he was quoted in 2003 as saying "we won" when referring to the results of the Intifada, years before it ended. He expressed criticism of Israeli miserliness to Abu Mazen during his term as Palestinian Prime Minister and maintained that a unilateral retreat from the Gaza Strip constituted a "tailwind for terror." In addition, he stated that Israel should release more Palestinian prisoners as a gesture to Abu Mazen.

The Chief of General Command also hinted a number of times that he opposed the over involvement of the Minister of Defense Shaul Mofaz in internal army affairs, as well as confronting the members of the Knesset Foreign Affairs and Defense Committee under MK Yuval Steinitz. Despite all of these issues, Yaalon was a central participant in the preparations for the disengagementå from the Gaza Strip.

Effects of Political-Military Relations on Military Resilience

Damage was caused to the *Military Framework (Cluster C)* in *ability to maintain the rules and procedures of a professional army(C1)*

The pressure imposed by the political echelon on the military echelon in an attempt to dictate the character of military operations, as, for example, the combat in Jenin, created conflicts of interest between the Chiefs of General Command, who understood the political pressure, and the forces in the field who, in order to carry out the mission in the correct manner were forced to fight slowly (with the understanding that combat in densely populated areas is slow by definition). Thus, in this case, the commanders refused to accept the directives and did not pressure the officers in the field, as they considered that if they had applied pressure to those in the field, it might have caused greater casualties among the combat soldiers.

In retrospect, the behavior of the army command in the field was proven to be effective in the test of combat results. But tension was introduced with the deviation from the orders of the senior army command, even though officers on the divisional and battalion levels blocked the pressure that emanated from the political echelon, and constituted a kind of filter for their directives. They did not transmit this pressure to the soldiers in the field.

Summary of the Effects of Political-Military Echelon Relations on Military Resilience

Consensus on the objectives and means that usually was the case in relations between the political and the military echelons led to a low level of damage to military resilience.

Israel–The Palestinian Authority: The Military Model, Warfare Strategy, Political-Military Relations and Military Resilience

Despite the limitations of immunity that apply to the publication of expressions of harm to the resilience of the IDF, it appears that the variable of warfare strategy had the greatest effect on military resilience. In other words, the greatest damage to resilience was during the first stages of the military conflict.

During the al-Aqsa Intifada, the army primarily made use of regular units, including units of border guards, for the most sensitive missions, especially in the critical opening phases of the military conflict, in constructing the significance of the Intifada in public discourse. Only at an advanced stage of Operation Defensive Shield were the reserves called up for important combat. Simultaneously, at the beginning stages, the army used combat methods that alienated the fighter from his victim, such as combat helicopters and tanks, the use of which minimized casualties among the combatants.[28]

The damage to military resilience from the military model including relations between the military and the political echelons was low. It appears that the close link between the military and the political in the Israeli reality cannot be ignored and this link characterizes these variables, each in itself and both of them together. Moreover, the two Chiefs of General Command at the time of the conflict saw themselves as public trustees and not only as government representatives and stressed this in their statements, which illustrated the lack of consensus regarding means. In other words, the behavior of the senior army command during the second Intifada embodied fusion or perhaps even confusion between *command* and *marketing*. While command demands taking risks, showing importunity vis-à-vis the operational forces, enforcing obedience, learning and internalizing lessons, among other requirements, marketing stresses presenting moves to consumers, the citizens, who are perhaps even the potential voters of tomorrow.

The description of damage to military resilience as a result of too great a burden on reserve soldiers can be attributed to the connection between the political and the military. The government and the army command failed to perceive the limitations in exploiting the reserve system. Thus, they overburdened the reserves in order to reinforce their aggressive stance in the territories about which there was consensus, until the Knesset reevaluated the amount of time permitted for annual reserve service.[29]

Table 6.1 Israel-The Palestinian Authority 2000–2004

Military Model, Political-military relations, Warfare strategy	*Description*	*Military Resilience*
Military Model	Integrated military models	Low Damage
Warfare Strategy	2000–2002 Low scale warfare	High damage
	2002–2004 Low scale warfare with a shift to large-scale warfare	High damage
	September 2004 Increase in large-scale warfare	Resilience recover
Politico-military relations	Generally consensus on aims and means	Low damage

Conclusions

Military Resilience in a Low-Intensity Conflict—France, Britain, Russia, and Israel

Comparing military resilience in the management of a low-intensity conflict between the states discussed indicates transversal trends that cross all and those that are unique to each state. An analysis of these trends reveals the effects of the military model, the warfare strategy and the relations between the political- military echelons in each state, on military resilience.

The effect of relations between political-military echelons on military resilience crosses borders. This is the most dominant variable in its effect on military resilience. In all of the states, these relations were revealed as causing damage to military resilience. The damage this causes is severe when it is characterized by an absence of consensus as to both the objective and to the means. However, even if and when consensus is lacking only as regards the means, it still brings about severe damage to resilience, as in the British case.

The strategy of small scale warfare also has a strong effect on military resilience in all the cases discussed. Although the measure of components in small low intensity warfare changes from state to state, whenever small scale warfare exists in the various arenas, there is serious damage to resilience. Two limitations to this finding should be mentioned:

1. The cases that have been discussed have shown that there is no warfare that can be characterized as solely small scale or conventional military warfare, as the evolution and transformation of the conflict in each state involved warfare that was a combination of both types, and in the Russian case, even unconventional means were also used.
2. In the Israeli case, the British case, and to a certain extent, also in the French case—a combination between two types of warfare strategies do

not allow their characterization as either type, but rather as a *unique professionalization* of low-intensity warfare. This professionalization cannot be attributed to the Russian case, as there, differentiation was maintained among the components of warfare on the three popular levels, that is, there were components of small scale warfare, conventional warfare and unconventional warfare.

Regarding *the model of the army: Here, too, there are effects that cross all cases in the sense of an ongoing process of transition of all of the armies and its implication on military resilience.* All of the armies are involved in a process of changing military model. In each of the states, this process brings about damage to military resilience when managing a low-intensity conflict. However, *in each state, a unique link was discovered between the given military model at the time of the conflict and military resilience*

The military model as a post-modern model that poses serious damage to military resilience can be seen in the case of Great Britain. The proximity of developments between the transition to a post-modern model in 1963 and the outbreak of the conflict in Northern Ireland in 1969 cannot be ignored. It may be assumed that this proximity adds to the magnitude of the damage to military resilience and reveals how this damage is expressed concretely in the field.

In Russia, as well, there is a similar tendency although the declaration of the political echelon regarding a transition to a post-modern army has not yet been formally implemented institutionally. The damage to military resilience in this case appeared to be low, but here too, the central factor involved in this damage should not be ignored. The damage suffered by the army occurred during a period of conflict against the backdrop of deep changes in the establishment, that Russia was experience in government-economy-society. It may be assumed that the actual transition, involving change in the model of the army to a post-modern one may have the same results as the British case.

In France, the model of a late modern army involves minimal damage to military resilience. During the period of conflict, a clear doctrine of transition to a post-modern army had not yet been formulated.

In the Israeli case, the mixture of different models has been found to have a low effect on damage to military resilience. This mixture expresses the same transition that is being experienced by the other states in this book, from which two opposing developments can be derived in the link between the military model and its resilience:

1. The existence of characteristics from a mixture of models indicates stagnation, on the one hand, and change on the other, which can be explained as

the lack of adaptive ability to systematically and chronologically undergo changes in the military model through time. In this case, the damage to military resilience may be higher with the mediation of complex transition.
2. The mixed military model evolves into a new, post-modern army, gradually and with considerations of doctrine and relevant security perceptions and not according to any particular format of models. This tendency is likely to enable more supervision and regulation of the connection between the military model and the geo-strategic, military social and economic realities and to prevent possible damage to military resilience.

This comparison of how the different states deal with the problem of military resilience can be summarized with the fact that this too crosses borders. France, Britain, Russia (only from the second campaign), and Israel have been shown to be concerned and involved in the erosion of military resilience in the extended conflict that they have conducted. Over and above the political aspect of this concern at home and abroad, all of the states have adopted similar measures to moderate the damage to military resilience. These measures have included:

- Changes in the framework of military model components, such as a change in the definition of missions, structural changes, technical changes and others.
- Adopting a system of propaganda to moderate the effects of the media and public opinion at home ad abroad.
- Censorship and emergency regulations.

In all states, the war for the support of public opinion is conducted with strict limitations on freedom of expression and strong censorship of the media. The demonization of the opponent and the negation of his/her membership in an appropriate civilian entity are conducted with the same means of psychological warfare.

A special emphasis should be made on the role of setting up virtual and ad-hoc borders, which are meant to separate the sides in each conflict, as derived from the decisions of the political echelon and from the cooperation between them and the military echelon. These are not politically recognized as autonomous borders, but as such that serve operational and tactical conduct of the conflict. Over and above the role of these barriers in continuing to manage the conflict on both strategic and tactical levels, a transition from the expressive dimension of warfare to its instrumental dimension is created, with the mediation of the political echelon. This transition moderates the damage to military resilience due to several reasons such as psychological aspects of territorial "them-us", differentiation between military and civilian etc.

> Structural, functional, and social-mental elements of the military model influence its resilience during a low-intensity conflict. The more the military model is post-modern, the greater the chances of damage to resilience in conducting a long-term low-intensity conflict.

In the case of Great Britain, we were witness to the great damage attributed to military resilience in a post-modern army in spite of the fact that relations between Society-Media-Military had not yet developed their post-modern character.

Damage to resilience caused by the model of the army constituted the starting point in the situation of the British army stationed in Ireland. The longer the conflict went on, the more the political echelons internalized the uniqueness of the warfare demands and supported the establishment of training programs that better suited field conditions. The weakness in the number of combattants and the "overstretching" of the army structure, which derived from the change to a small professional army, was halted by rewording a written military doctrine that was now composed especially to manage a low-intensity conflict. This contributed to an improvement in the quality of military resilience and to coping with the problem of the reduction in the size of the army.

The validity of this insight is derived from two types of findings:

1. Relations between society and the army and between the media and the army in Britain were not found to have significantly damaged military resilience, as, at the height of the conflict, both relations with society and the media became more like components of the modern and the late modern military models even though, formally, Britain had adopted the post-modern model.
2. The findings became clear when dealing with the other states in this book. In France, in Russia and in Israel, the post-modern model has not yet been adopted. With the exception of France, Russia and Israel have weighed the possibility of adopting a post-modern model. In these three states, which employed the modern military model, and the late modern model, in combination with one element or another of the post-modern army, damage to military resilience was found to be low.

In an analysis of characteristics of the military model in various states, it was found that in all of these states, one element or more of the post-modern army could be found, while *society-military relations and media-military relations* serve as a form of control for fluctuations of the post-military model variable. In all of the states, *society-military relations and media-military relations* were not post-modern at the time of the conflict. These are dynamic, rather than static elements that constitute a kind of sensor for checking the resilience of the army.

Does the Post-Modern Military Model Always Constitute a Danger to Military Resilience?

The transfer from the modern military model to the post-modern model involves a possible upheaval in the reciprocal relations between the political, the social and the military systems and to what is required from these relations in managing a continuing conflict. The timing of the change from a modern army to a post-modern army, while managing a low-intensity conflict or just before such a conflict that is immediately expected, may create elements of friction in carrying out the change. The political and the military echelons must take the timing of the change to a post-modern model into account considering its effect on military resilience.

In this context, a consensus between the echelons regarding the change from one model to another must exist. This condition is problematic against the background of the possible blurring of borders between the political and the military stemming from the change in the nature of war and in the military model in the present era. There is a need for adapting flexible implementation of post-modern elements with strategic and tactical modifications that become necessary due to the timing and momentum of the shift to another model.

Regarding conscientious objection:

- Institutionalized conscientious objection does not necessarily predict damage to military resilience.
- This finding is dependent on two conditions:

1. When conscientious objection refers to refusal to be drafted to compulsory service for reasons of conscience.
2. When this refusal is permitted and is anchored in law, enabling the substitution of civil or national service to being drafted.

These conditions express the establishment's recognition of pacifism as a legitimate individual right, and simultaneously, the tie between citizen and state is maintained by national civil service. In other words, the state participates in the implementation of this value and grants it to the citizen by right, as this expresses equality under the law for the individual who serves and the individual who does not serve in the opinion of the state (excluding incentives and remunerations that the state uses to compensate those that serve in the army).

When conscientious objection is not anchored in the law and exists in practice, even if it occurs only in isolated cases, this negates the link between implementation of a value and ties to the state. It is done in opposition to the will of the state and is accompanied by imprisonment. This is actually a negation of recognition of state authority against the backdrop of the rational perception of the citizen, and in its place, an aggressive, manipulative and compulsive perception of this authority is created. This perception may reflect upon the justice of the state within the framework of the conflict. A conscientious objector may attribute his/her behavior to his/her individual viewpoint and, at the same time, recognize the different viewpoint of the citizen who serves in the army. However, the negation of his/her right may bring him/her to non-recognition of the rights of the other. On the other hand, a dispute about the concept of "rights" when it concerns the army may damage military resilience.

Regarding conscientious objection *during military service*, this has been found to damage military resilience. Checking the effect of conscientious objection on military resilience, on one hand, and conscientious objection as an expression of damage to military resilience, on the other, does not create a tautology. This is due to the following reasons:

1. The question of conscientious objection as an element in a military model depends on the laws of the state. This has been discovered not to result in damage to military resilience.
2. The question of conscientious objection that arises during military service does constitute "denunciation from within" the military system and is seemingly based on the experience of military realities. The soldier who has been inducted began his army service with great motivation and with the will to make an effort to contribute to the state. The dramatic change from being highly motivated to becoming a conscientious objector expresses dissonance and disappointment with army service even if the disappointment refers to one specific operation only.
3. This refusal to serve may encourage pre-service conscientious objection.

The conclusions regarding the relationship between the military model and resilience are not discussed in the element of threat perception. This element, although it is also characterized by subjectivity, relativity and dynamism, can describe the "atmosphere of the given situation" between the state and the hostile force, when all other components of the model are in the internal-political realm of the state. The role of this element is in clarifying the book's paradox that has been described. Its significance in the problem of resilience demands the introduction of an additional concept called "threat attribution" that may solve this paradox.

Discussion and conclusions regarding the military model raise a general question that cannot be ignored. The post-modern model primarily characterizes only one state: Great Britain. That being the case, what is the status of a theory based on only one single example when several other states are compared? It would appear that the fact that the armies of the other countries include a combination of characteristics, some of which are features of the post-modern model, provides an answer to this question. While the models of the other countries are indeed not classified as post-modern models in all of their characteristics, but include certain features of this model that have had some effect attributed to the post-modern model, this reinforces the theory of the effects of a post-modern model on military resilience.

The strategy of warfare in a low-intensity conflict affects the resilience of the army:

> *The more distant warfare strategy is from the army's designated mission and the formal military training of its military system, the greater the chances for damage to military resilience. The more the warfare in the conflict uses small-scale military tactics, the greater the chances for damage to military resilience.*

France, Great Britain and Israel have conducted their conflicts using more features of small scale warfare than Russia. The Russians entered Grozny using regular military tactics, despite the built-up urban area that limited this type of warfare. However, in all of the states examined, military resilience suffered high damage.

There is nothing in the Russian experience that negates the link between military strategy and military resilience. However, this link can be limited to the following conditions:

- In an asymmetric conflict between a political entity and terrorists/ guerrillas, which from the beginning includes characteristics of conventional military warfare (developing or not developing into unconventional warfare, as seen in the case of Russia) the chances for damage to military resilience are great and irreversible.

- However, if, from the beginning, there are more features of small scale warfare that develop into conventional military warfare, the chances for damage to military resilience are high only at the beginning of the conflict and the damage is reversible. The damage to military resilience lessens as the conflict becomes more conventional and evolves to regular warfare.

An explanation for this can be seen in the consent to use more military power when, due to the inferiority and asymmetry of the other side, its limited use has not prevented great and sensitive damage to life and property. This idea invites reference to the connection between learning in the military system, and its resilience. Put differently, despite the sensitive character of the low-intensity conflict and the military, political and ethical dilemmas involved, participants in fighting of this type learn to internalize the mutual relations between the nature of their activity and their resilience as an army. When they discover that not only does low-intensity warfare not achieve the objectives of the conflict and damages their resilience but also, by its very nature, it escalates to dimensions of strategic damage to life and property, different field conditions are created that lead to consideration of more massive military options. At this stage, there is realization of the "necessary evil," with no frustration, guilt, or loss of sense of mission, but rather deriving from the lack of relativity between the investment necessary to conduct low-intensity warfare and the failing results in the field. In this case, the military system learns and internalizes from experience that the threat requires and justifies the use of conventional military warfare. This evolution gets stronger when the opponent, that has been inferior in terms of capabilities, reduces the asymmetrical nature of the conflict.

In the Russian-Chechnyan case, it can be said that the Russian troops in Grozny were not given the opportunity to internalize this learning, due to the fact that they began right away with conventional military activity.

Another conclusion regarding the connection between the development of a conflict, learning and military resilience is connected to the type of threat being faced and the warfare strategy:

- The outbreak of warfare with a terrorist/guerilla organization is derived from the type of threat it presents to the political entity and from the length of time the threat exists. A tactical threat in a limited measure of time, managed as a low-intensity conflict, will not damage military resilience.
- A tactical threat for a long period of time will require the use of conventional warfare means, as well, in order to protect against erosion of military resilience.

- A tactical threat that has become strategic will require the use of conventional warfare, while the learning referred to above should protect military resilience.

The hypothesis regarding the connection between warfare strategy and military resilience leads to the development of an interesting conclusion about the role of the political-military echelon in this hypothesis. Raising the level of warfare to conventional military strategy and even a particularly intensive level makes it possible to protect military resilience on condition that this is in line with the objectives and instructions of the political echelon. That is to say, using more fire power or more military victories will not ensure resilience unless there is congruence between these and the objectives and instruction of the political echelon. There is clear proof of this in the French and the Russian cases. Whenever the political echelon announced political negotiation with the rebels- the military echelon raised its threshold of operational activity. The sense of being on the verge of victory and defeating the enemy was sharpened in the military echelon as proof of its power to manage the conflict. In the two cases, increased motivation to use intensive means of warfare did not contribute to military resilience. The opposite was true, as the political echelon had not had a hand in the military activity.

Relations between the Political-Military Echelons in Managing a Low-intensity Conflict Affecting Military Resilience

> *The absence of consensus between the political and military echelons regarding objectives of the conflict and/or the means to manage it raises the chance of damage to military resilience.*

This hypothesis is greatly strengthened in this book and is expressed by the fact that even when the absence of consensus is only on the level of means, as in the British case, we viewed great damage to military resilience. This is all the more true when there is no consensus regarding the objectives of the conflict.

Another expression of the strength of this hypothesis derives from the conclusion that damage to military resilience is not necessarily connected to the lengthy time dimension of the low-intensity conflict. This conflict has a characteristic opening point that will cause a weakening of resilience even at the beginning of the conflict. The absence of consensus between the two echelons in the British and the Russian case proved this. In the Russian case, relations between the political and military echelons present an empiric dimension, "cleansed" of the effects of the alternative variables, which theoretically could have been those that affected military resilience. The effects of the absence of consensus was already prominent in these relations from the beginning of the campaign in Grozny.

When we distinguish between damage to resilience at the military command level and damage at the level of the junior soldier, this hypothesis needs examination in greater depth. The chances of damage on the senior military command level are greater against the backdrop of no consensus between the political and military echelons. Although there are ramifications for the junior soldier, as well, the senior military command is closer and more accessible to the political echelon. The case of the putsch in France and the refusal to carry out orders of the political echelon in Russia illustrate this point.

The strength of this hypothesis is also reinforced by the link between political-military echelon relationships, and the rehabilitation of military resilience that has been damaged or is in danger of being damaged.

a. In the connection between military model and its resilience, the political echelon plays a central role. It is responsible for the design of the model, and its integration into the doctrine, and the updated security perceptions including all of their components. Regarding the elements of relations between the military and society, and between the military and the communications media, as well, the political echelon plays a central role in the framework of relations between the center and the periphery.
b. In the connection between warfare strategy and military resilience, the political echelon also plays a central role. The military strategy is derived from the strategic political goals and/or the broad strategy. Here, too, the role of the system of mutual relations between the two echelons is clear.

The logical conclusion is that all rehabilitation and preservation of military resilience is dependent on these relations. In a situation in which this resilience suffers a blow from the military model and/or from warfare strategy, rehabilitation is almost impossible if there is no consensus between the two echelons as to objectives and means to conduct a low-intensity conflict.

A relevant question is how to identify various trends linked to military resilience taking place at the same time. Specifically, how can it be argued that military resilience is damaged by the use of small-scale warfare strategies, when, at the same time, resilience is maintained by the existence of consensus between the political-military echelons (as, for example, in the French and Israeli cases). An answer to this question requires a division, although fairly artificial, between the senior military command and the junior command. Thus, it may be argued that warfare strategy has been found to particularly affect the resilience of the warfare systems in practice, that is, the soldiers who are involved in warfare and who are witnesses to its results, rather than on those who have planned the strategy. (Quite often, this aspect of the damage to military resilience is related by the senior military echelon to the political echelon, for example, in a request for a new evaluation of more suitable warfare methods and means to rehabilitate the morale of their soldiers.) On the other hand, the senior military command, which is naturally closer, more sensitive and especially attentive to the political echelon, is affected by its relations with this echelon. Thus, it is certainly possible to have a situation in which military resilience on the level of senior military command is maintained while the resilience of the junior military command is damaged. On the other hand, damage to resilience on the level of the senior military command will necessarily damage the resilience of the junior military command.

This poses a unique challenge to the senior command to remain especially aware of and to safeguard military resilience. It should be added that it is important for the operational warfare system in the field to have the support of society, but it is even more important for it to have the support of the senior military command, itself. A clear expression of this support is in strategic, tactical and logistic planning of the military presence in the field that will create an atmosphere of defense and security for the troops that is necessary both for their presence in the field and for managing warfare under conditions that are not dependent on them.

Another conclusion is connected to the mission of the troops and to their training. To the extent that operational warfare is not conducted according to the military mission, the chances of damage to resilience will increase. Being drawn perceptually into the nature of the asymmetry between the sides in a conflict may play into the hands of the terrorist organization regarding warfare strategy. Military warfare, characteristic of an army that utilizes field conditions to the fullest, with fewer tanks in an area of civilian population, more professional commando and air force units, improvements in intelligence abilities, and destruction of objectives from afar, not only supplies a strategic advantage, but also has the effect of safeguarding resilience, that is, a reduction in the expressive element of warfare and an emphasis on the instrumental element enables warfare according to the designated role of a military and safeguards resilience.

The problem is that "drawing" the opponent into conventional military warfare may elicit moral dilemmas with severe effects on military resilience. On the other hand, delimiting the opponent in geographic terms may translate the warfare advantage into adoption of rules of the game that are in line with the conventional military warfare mission. The limits, borders and even virtual barriers placed by the armies of the states discussed involved a number of consequences:

- Dealing with the problem of mixing civilian and military aspects and its effect on the feelings of the citizen-soldier.
- Constructing the rules of the game of deterrence between the sides and creation of a "fairer" environment for using conventional military force, if deterrence fails.

Recommendations in the Realm of the Connection between the Military Model and Military Resilience

A solution to the paradox presented in this book is possible employing a number of practical actions:

In states that already formally have a post-modern army (as in Great Britain): An attribution threat should be formulated in addition to the threat perception characteristic of a post-modern army. While the latter constitutes primarily the most logical or the most serious threat possible, the former is the principal threat to which the decision to respond is necessary. This does not mean that integration of threat perception should be ignored, but rather that the first priority should be given to building a force to deal with the attribution threat. In other words, the direction of technological and professional development that characterizes the post-modern army must consider the attribution threat. In an era of global terrorism, the concept of threat as "internal crisis" reduces the complexity of the challenge that this threat presents to a post-modern army.

Thus, force building and advanced technical improvement that are suitable in a low-intensity arena cannot constitute only an adjunct to the doctrine of future warfare arenas with various states and/or with strategic military threats from far away countries. Allocating importance to managing a low-intensity conflict is necessary from the general strategic level to the tactical level. This must be expressed in top quality technologies and training that is directed at low-intensity conflicts.

The present research was impressed with the importance of psychological warfare and with military intelligence that has accompanied the conflict arena in all of the states covered by the study. Although this book does not discuss these two issues, preference to technical advances in these areas is of prime importance in actual strategic policy.

It can be assumed that threat perception of inner crisis and/or a non-military threat as an attribution threat, along with more comprehensive threat perception, in combination with diverting technology to this threat, may solve the paradox between threat perception of an inner crisis and post-modern technology.

States in which formal adoption of the post-modern military model has not yet taken place (Israel, for example): Every decision of the political and military echelons that should signal a formal move to a post-modern army (for example, the decision to change conscription patterns, length of service and others) should take the attribution threat into consideration, in addition to threat perception and the necessary resulting priorities, as detailed in the last clause.

Israel has been described in this book as a case of complex threat perception, and thus, the distinction between an immediate threat and a long term threat demands continual updating against the backdrop of the dynamic link between these threats. This link is expressed in the quantity and quality of possible fronts included in each threat perception.

Thus, the response that has been provided in the conclusions to the question of whether the post-modern army model always constitutes a danger to military resilience, applies here, as well. In a case like the Israeli one, in which the attribution threat and threat perception are integrated, the fact that these are dynamic and that they constitute options for complex fronts poses a challenge for great operational-quantitative-technological flexibility on military resilience.

Regarding the change of the military model to a post-modern one that is "more friendly" to military resilience, so that the relations between society and the military, communications and the military, and conscientious objection do not undermine resilience, deriving concrete recommendations poses a dilemma. Over and above the well-known recommendations regarding the need for diplomatic information services, military and other means of advancement, and improvement of public opinion at home and abroad, and for the army involved in a low-intensity conflict, these must depend on the development of structural processes and different mutual relation than those that have existed in the civilian and military spheres until then.

It is sufficient to point out elements in the military model that are dependent on decisions of the political-military echelons, and others that are not dependent upon decision-making. For example, conscription is an issue of decision-making, although the decision is usually accompanied by previous cultural-political-social changes that, to a great extent, have prepared the timing for the decision indicating that there is a need for the change. On the other hand, relations between the society and the military and/or relations between the media and the military, which also affect the development of the decision to change the form of conscription, do not require a decision on the part of the political-military echelons in order to create a post-modern or any other character. These are long-term structural changes. Thus, drawing conclusions about involvement in these changes is possible, but extremely limited.

Regarding conscientious objection, the conclusions support the formal establishment of conscientious objection. This should not be avoided if the act is anchored in the legal authorization of the state, and army service is converted into the possibility of some type of required national service. On the other hand, conscientious objection that occurs during army service should not be permitted and should be accompanied by sanctions. If it is "condemned from within" by the conduct of the military system, this might undermine resilience on the framework and in the unit levels.

Recommendations in the Realm of the Connection between Warfare Strategy and Military Resilience

Recommendations derived from the conclusions that have been drawn about the link between warfare strategy and military resilience, first of all, caution against both a perceptual and a functional failure. This refers to a situation in which there is too great a reliance on technological and training developments as regulators of a low-intensity conflict, which, as stated, are decisive in their importance. This may create a process of continual adaptation to the rules of the game set by the non-conventional side, which may damage military resilience. Thus, a reversal of these adaptation processes should be aimed for so that there is gradual movement towards conventional warfare and a regular army mission, that is, being aware of the erosion created by the process of adaptation and trying to stabilize warfare with the classical relative advantage of armies. Therefore, the opposing side should be drawn towards this arena wisely and its steps should be controlled with supervision and warfare from a distance.

The creation of a delimited geographical area with separation lines, fences and/or virtual borders supplies convenient geo-strategic conditions for warfare from a distance. Above and beyond the importance of this delineation for separating the civilian and the military, "status quo lines" are created, which, although they are fluid, if they are breached by the unconventional force, may be accompanied by military warfare from afar. Although the creation of geographic delimitation may begin as tactical thinking only, it is etched into the consciousness of the sides and evolves into the rules of the game based on deterrence. Geographical delineation enables preparation for warfare from a distance in the case of provocation by the unconventional side, and may create less "noise" at home and abroad.

This is the place to recommend thinking and perceiving of conflict not in order to derive concrete rules of action but rather as a call for awareness on the part of the political and military echelons of conflict development, integrating the asymmetry of the limited conflict that should then turn into a conflict between equal political entities. How does this occur?

Posing separation lines and geographical delimitation, despite supervision and warfare from a distance, may lessen control over massive armament of the organization that up to now was unconventional, and it may be trained for conventional military warfare, to all intents and purposes. Continuing to manage the conflict with military and/or political means will not prevent this armament. However, the move to symmetry between the sides may enable its management as a conflict, not necessarily making use of warfare at all and certainly not with the elements of small scale warfare. This is because the symmetry constitutes a structural and functional basis for building deterrence.

This stage enables the beginning or the continuation of a political solution to the conflict simultaneous to continuation of military activity whose aim is, among others, stabilization of the rules of the game of deterrence between the sides.

Despite the importance of conventional warfare to maintain military resilience and the military mission, the book does not recommend adopting this form of warfare at the beginning of a low-intensity conflict. In a low-intensity conflict the army has no choice but to learn "the hard way," that is, to adopt features of conventional warfare after the failure of features of small scale warfare. If it attempts to adopt conventional warfare from the beginning, not only will the army fail to maintain its resilience, but the damage caused will be irreversible. Therefore, there is no recommendation for sparing the army the necessity of learning "the hard way" and enabling it to use conventional warfare from the beginning, perceiving it as "a necessary evil" and preventing damage to army resilience. There are two salient reasons for this:

1. The perception of "a necessary evil" is limited in its construction if it is achieved only by rational persuasion, and it is more stable in its formulation when it results from the conscious experience of the fighter. The fighters must have the direct experience, accompanied by learning about the heritage of relevant battles that indicate the need for the perception of the "necessary evil." They must be convinced that the warfare methods being used, either by them or by those who came before them, do not hold up under the test of results. They must internalize that the justice of their adoption of conventional warfare in the conflict is not a result of cold rational consideration, but rather of a necessary development verified in the field. The justice of the turn to conventional warfare comes into being gradually with the decrease in breathing space necessary for managing a low-intensity conflict.
2. The perception of conventional warfare as a solution considered "a necessary evil" must find resonance at home and abroad. In these arenas, the conflict develops in its political, social, cultural and economic aspects and the rules of what is permitted and what is forbidden become more distinct in defining consensus and lack of consensus on the level of means of man-

aging a conflict. Here, the use of conventional warfare may find appropriate resonance and be less "noisy."

In actuality, the recommendation that is apparently indicated of letting the army "learn the hard way" does not exist in reality in the sense of some kind of preliminary decision on the part of the political-military echelons who know what is coming, identify a low-intensity conflict in advance, predict its development and scenario, and then determine that type of warfare to employ. The social-civilian character of a low-intensity conflict that, at times, breaks out as disorderly social conduct, demonstrations and the like, and develops into armed conflict, testifies to an unexpected process of interaction between opponents. Thus, the lack of ability to recommend using conventional warfare of one measure or another at the beginning of the conflict is a warning of the complexity of such a threat, of the lack of certainty that it creates, against the backdrop of an inability to catch up with or to control the rate of its transitions, and especially of its being a continuing process. Only with a certain momentum in this process will conventional warfare be a fitting response. Every conflict discussed in the present research has had its own momentum. Any clear recommendation that will propose shortening the process of learning may result in an analogical failure.

Recommendations in the Realm of Connections Between Political-Military Echelon Relations and Military Resilience

In an era of warfare against terrorism and in an era of post-modern military technology, in which the demarcation between political and military strategy is unclear on the functional level, structures must be set up to regulate and supervise the conflict between the political and the military echelons. This conflict stems from an excess of cooperation created by the nature of the conflict on one hand, and from intervention by the political-civilian into the military, on the other, which may become more moderate with the encouragement of administrative bodies, such as a National Security Council.

Although the effectiveness of administrative bodies depends to a great extent on the political culture of the senior echelons themselves who determine the extent to which they are advised by and consider these bodies, it is important at least to make sure that there are ways of managing the tension between the two echelons. The political and military echelons should internalize the effects of the tension between them on the warfare systems in the field, and manage this tension with the required restraint and responsibility. In a situation in which the tension between the two echelons gets out of control, the basis of the "raison d'etre" of the conflict may collapse on the battlefield itself.

The discussion of military resilience in this book for the first time raises connections using new variables that affect resilience in managing a low-intensity conflict far beyond military psychological and organizational aspects. Bringing these connections up for discussion contributes to:

- The identification of ways of coping to maintain resilience in armies that today characterize many states in the global arena.
- Understanding a low-intensity conflict with a terrorist/ guerilla group, not only on a theoretical-conceptual level but also, and principally, in the nature of managing military warfare against such a group.
- Checking the relevance of reorganization of the Israeli army especially, as it is still managing a low-intensity conflict, and the potential for a conventional war still exists.
- Raising the profound dilemma: Should an army adopt changes in its warfare theory, warfare means, and unique preparations, in order to serve the asymmetric principle, or should it force the opponent into an arena that will maintain the advantage of military action for which the army was intended for, to begin with. This is a dilemma that has implications not only for the problem of resilience, but also for the whole system of opponent perception.
- Understanding the asymmetry between the sides carrying on a long-term low-intensity conflict and internalizing this insight in the framework of political processes to reduce the conflict.

This book does not deal with additional variables that may possible affect military resilience or, in addition, expressions of damage to military resilience. For example: *The relevance of the socio-economic, demographic and ideological status of those involved in warfare.*

The military warfare system discussed in this book poses not only an issue of military identity. It involves soldiers with their own socio-economic and political identities, among others, that is, citizens of the state. This is a factor that, in advance, limits the discussion of military resilience when dealt with in the military framework only. In reality, this is an artificial isolation, separating between the military and the civilian in an entity of fighting soldiers. In other words, this book assumes that social-civilian-political-economic and other characteristics also affect the resilience of the army.

A concluding rationale that needs further research concerns the four societies in the four cases this book has discussed. It concerns the evaluation of the relationship between *military resilience, national resilience and social resilience*. Military resilience is derived from the national resilience of the state. However, unlike military resilience, national resilience reflects more on social resilience and expands to a vast realm of aspects with direct relevance to society. The evaluation of this relationship can help in assessing interesting insights on the four societies of the four cases discussed in this book.

The relationship between military resilience-national resilience-social resilience differs between the different countries. Paradoxically, the more the military is post-modern, the less correlation between the military—national-social resilience. The less the military is post-modern, the more correlation between the military-national-social resilience. The national aspects of armies are weak in post modern militaries, and the book has elicited this weakness in different aspects. Though none of the armies in the countries discussed has totally developed into a post-modern military, some cautious conclusions could be tentatively derived: The weakest correlation between military-national-social resilience is better found in Britain, while the strongest correlation is best found in Israel. France and Russia are characterized with complex correlations. Yet it reflects the discussion on the various social aspects the book has discussed within the military resilience of each country.

- The heritage of previous conflicts, that is, the contribution of the memory of previous low-intensity conflicts by the former generation, testimonies of soldiers from the field and others, whether this occurs within the intimate family, collectively or in the media, affects the legitimacy of the present conflict. Raising memories of a previous conflict, which are charged with personal emotional residue against the establishment and against the management of the past conflict may damage military resilience.
- The geographic distance between the home country and the conflict arena and its effect on army resilience in general and on those involved in warfare in the field, in particular.
- The settlers, an issue that exists in every conflict. The connection between the culture of the opponents and army resilience, especially in light of the fact that in every conflict there are settlers who are citizens of the state that is a side in the conflict.
- The element of psychological warfare, a profile of the military information system and the conduct of the senior command as the guardians of resilience.
- Others.

It would be positive to designate research that is socio-political in character, to check the characteristics of the conception of a dispute as viewed by the political echelon, society and the army, and the mutual relations between them. This would unite not only the issues of granting legitimacy, a war of choice or not, a values-material-ideological conflict, but primarily it would design the way it is managed. In addition, there could be a comparison of the two sides, that is, including the conceptions of fighters belonging to terror and guerilla organizations. In this way, it would be possible to identify the characteristics of differentiation in the mind and conception of a dispute, among the players in the state (the political echelon, society, the army) and in the society being created, as expressed, for example, in the potential of the struggle of a terrorist organization for sovereignty and independence.

Regarding the present research, there is room for research that would try to identify *what were/are the components of resilience for the other side in conflicts* discussed. In other words, what are the components of resilience for the Algerians, the Irish, the Chechens and the Palestinians.

In addition, there would be place for a comparative research between perception of military resilience of the senior command and that of the junior command. This research could supply valuable insights into application of treatment strategies for safeguarding military resilience.

A future study could contribute to the issue of organization learning accumulated by the army in a low-intensity conflict to strengthen its resilience. This research could contribute to understanding the conditions this learning creates and the way it supports military resilience.

The present study has discussed the role of artificial and virtual barriers and boundaries that the armies has succeeded in setting up between themselves and the other side. This subject is also worth researching in depth to understand the change in the rules of the game between the sides in managing the continuation of the conflict with military and/or political means.

All these themes should be investigated on al forms of dynamic low-intensity conflicts when the most actual one taking place right now when this book is written is the conflict in Afghanistan along Global and virtual terrorism. According to the insights proposed by this book, these conflicts embody "unripened conflicts."

Notes

INTRODUCTION

1. Westman, M. (1990). The Relationship between Stress and Performance: The Moderating Effect of Hardiness. *Human Performance, 3*, 141–155.

2. Kobasa, S. (1982). Commitment and Coping in Stress Resistance. *Journal of Personality and Social Psychology, 42*, 707–717.

3. Lazarus, R.S. (1996). *Psychological Stress and The Coping Process*. New York: McGraw-Hill.

CHAPTER 1

1. Stouffer S. A., *The American Soldier,* Princeton, NJ: Princeton University Press, 1949.

2. Britt T. W. and Adler A. B., (eds.). *The Psychology of the Peacekeepers,* Westport, CT: Praeger, 2003.

3. Kearney E. G., Creamer M., Marshall R. and Goyne A. (eds.). *Military Stress and Performance,* Melbourne Univ. Press 2003.

4. Meichenbaum D. and Jaremko M. E. *Stress Reduction and Prevention,* Plenum Publishing Corporation 1984.

5. Such as papers of the Behavioral Sciences Department of the Israeli Defense Forces, 2003.

6. Bartone, P. T., *Psychosocial stressors in future military operations.* Paper presented at the Cantigny Conference Series on Future of Armed Conflict, Wheaton, IL 2001; Bartone, P. T., Adler, A. B., & Vaitkus, M. A. "Dimensions of psychological stress in peacekeeping operations" *Military Medicine,* 1998 *163,* 587–593.

7. Bartones (2006), pp. 3–6.

8. Sarkesian, Sam C. and Scully, William L (eds.), *U.S. Policy and Low-IntensityConflict*, New Brunswick: Transaction Books, 1981, pp. 20–21 in Schultz, G. "Low

Intensity Warfare: The Challenge of Ambiguity" *U.S. Department of State Current Policy*, No.783, Jan.1986; Harkabi, 1990.

9. Calwell, C. E. *Small Wars and Their Principles*, Univ. of Nebraska Press, 1996.

10. Golan H. and Shay S., (eds.). *The Limited Conflict*, Tel-Aviv, 2004.

11. Field Circular "Low-Intensity Conflict," FC 100–20, U.S. Army Command and General Staff College, Fort Leavenworth/Kansas, May 30, 1986.

12. William, J. A. "The Post Modern Military Reconsidered," in Moskos, C., Segal et al (eds.). The Postmodern Military, NY: Oxford University Press, 2000; Kennedy, P. *The Rise and Fall of the Great Powers*, Davar Publications, 1992 (Translated to Hebrew); Cohen, S. (ed.). *Democratic Societies and their Armed Forces*, London: BESA Studies in International Security, Frank Cass Publishers, 2000.

13. Moskos C. in Cohen, S. (ed.). *Democratic Societies and their Armed Forces* London: BESA Studies in International Security, Frank Cass Publishers, 2000, pp. 3–22.

14. Ibid.

15. Harkabi, Y. *War and Strategy* (Ministry of Defence, 1992) (in Hebrew)

16. Ibid., p. 49.

17. IDF—Behavioural Science Department January 2003 FV 70–10.

18. Cited in IDF—Behavioural Science Department (ibid.).

19. Ibid.

20. Harkabi p, ibid., p. 510.

CHAPTER 2

1. Harkabi 1990, pp.337–338.

2. Based on Moskos 2000, ibid.

3. Based on the table in Harkabi's book 1990, p.125.

4. The book will deal only with the local aspect of media-army relations; that is, the discussion of international communications and media will not be discussed.

5. This refers to refusal to be *inducted into military service* because of conscience and/or political reasons. This should be distinguished from refusal due to conscience and/or political considerations *during military service* of those who have already been recruited and are serving in the army—that is, "functional refusal." This aspect is in itself seen as a damage to military resilience.

6. Moskos 2000, p. 8.

7. Harkabi, ibid.,pp.122–123.

8. Harkabi, ibid.

9. This issue will not be discussed, although this aspect cannot be ignored.

CHAPTER 3

1. Jacques D. and Ravid G, *Models of Foreign Armies*, A Social-Military project, 2001, pp. 119–141.
2. Ibid.
3. Charters, David and Tugwell, Maurice (ed.). *Armies in Low-Intensity Conflict–A Comparative Analysis,* London: Brassey's, 1987, pp 102–107.
4. Jacques and Ravid, ibid.
5. Charters and Tugwell, ibid., p. 107.
6. Jacques and Ravid, ibid.
7. Alexander M.S., Evans M., Keiger J.F. (eds.) *Algerian War and the French Army, 1954–62,* 2002 (translated to Hebrew in 2004), pp 275–278.
8. Martin L., "From Algiers to N'Djamena: France's Adaptation to Low-Intensity Wars, 1830–1987," in Charters, David and Tugwell, Maurice, ibid., pp. 96–97.
9. Alexander, ibid., p. 39.
10. Jacques and Ravid, ibid., p. 141; Alexander, ibid., p. 262.
11. Algerian War Readings www.usfca.edu/fac-staff/webberm/algeria.htm; Merom G., *How Democracies Lose Small Wars: State, Society, and the Failures of France in Algeria, Israel in Lebanon, and the United States in Vietnam*, Cambridge University Press, 2003, pp. 99–120.
12. Merom, ibid.
13. Jaques and Ravid, ibid.; Merom, ibid.
14. Author's translation, Alexander, ibid., pp. 56–57.
15. Ibid., p.138.
16. Ibid.
17. Ibid., p. 261.
18. Horen 1989, p.280.
19. Merom, ibid., p. 114.
20. Alexander 2004 p. 56.
21. Ibid.
22. Merom, ibid. p. 112.
23. Alexander, ibid., pp.251–264.
24. Alexander, ibid., p. 32.
25. Merom, ibid. p. 86.
26. Alexander, ibid., pp. 268–272.
27. Alexander, ibid., p. 122.
28. Alexander, ibid., p. 122.
29. Shrader C. R., *The First Helicopter War: Logistics and Mobility in Algeria, 1954–1962*, Praeger Publishers, 1999, pp.38–48.
30. Shrader, ibid.
31. Shrader, ibid., p. 139–140.
32. Shrader, ibid., p. 32.
33. Alexander, ibid.

34. From the testimony of a senior commander in the Legion units, Alexander, ibid., p. 20.
35. Alexander, ibid., pp. 9–13.
36. Chater and Tugwell, ibid., 1987, pp. 9–13.
37. Alexander, ibid., pp. 79–120.
38. Merom, ibid., pp. 83–95, pp. 258–260.
39. Martin, ibid., p.96.
40. Alexander, ibid., pp. 124–125.
41. Martin, ibid., pp 96–98.
42. Martin, ibid., pp. 86–97.
43. Alexander, ibid., pp. 131–138.
44. Merom, ibid., pp. 84–85.
45. Alexander, ibid.

CHAPTER 4

1. Jacques & Ravid, ibid., 2001, pp. 207–234.
2. Ibid.
3. Ibid.
4. Dewar, M. *The British Army in Northern Ireland,* Cassell Group Wellington House, 1985, pp. 30–32.
5. Ibid.
6. Ibid., pp. 58–60.
7. Ibid., pp. 58–60.
8. Jaques and Ravid, ibid., pp. 209–210.
9. Ibid.
10. Ibid., pp. 216–218.
11. Strachan, H. *The Politics of the British Army,* Oxford: Clarendon Press, 1997, pp. 183–184.
12. Ibid.
13. Dewar, ibid., pp. 32–226.
14. Ibid., p. 227.
15. Ibid., pp. 49–51.
16. Ibid., pp. 177–205; Charters 1987, ibid., p. 228.
17. Ibid., pp. 177–206.
18. Impressions British Soldiers Speak Out: cain.ulst.ac.uk/othelem/people/accounts/Lindsay.htm.
19. Dewar, ibid., pp. 32–48.
20. cain.ulst.ac.uk/othelem/people/accounts/Lindsay.htm.
21. Dewar, ibid.
22. Ibid.

23. Ibid., pp.39–47.
24. Ibid., pp. 180–185.
25. Ibid.
26. Ibid., p. 197.
27. Ibid.
28. Ibid., pp. 177–205.
29. MOUT-k48-www.geocities.com/Pentagon/6453/nireland.html.
30. Strachan 1997, ibid., pp. 181–183.

CHAPTER 5

1. Tazpit, International Military Thinking in World Armies Nov. 1996, pp. 83–88 (in Hebrew).
2. Bowker, M. & Cameron R., (eds.). *Russia After the Cold War* (Pearson Education Limited, England, 2000 pp. 223–225).
3. Ibid., 224–242; Tazpit, ibid.
4. Tazpit, ibid.
5. Bowker, ibid., p. 227.
6. Ibid., 228.
7. Historical Spotlight, ZARKOR April 2001, pp. 7–8 (in Hebrew).
8. Barylski, R. V. *The Soldier in Russian Politics.* New Brunswick: Transaction Publishers, 1998, p. 328.
9. Historical Spotlight, ibid.
10. Barylski, ibid. pp. 300–320.
11. Ibid.
12. Historical Spotlight, ZARKOR, 5, 2001, p. 17 (in Hebrew).
13. Barylski, ibid.
14. Historical Spotlight 2001, ibid., p.18.
15. Barylski, ibid., pp. 322–323.
16. Bluth, C., *The Collapse of Soviet Military Power* Darmouth Publishing Company, England 1995.
17. Spotlight: IDF Thinking, Nov. 2000, 4, p. 35 (in Hebrew).
18. Ibid., pp. 27–48.
19. Historical Spotlight 2001, ibid., p. 11.
20. Ibid., pp. 47–48.
21. Historical Spotlight 2000, ibid., 4, pp. 23–25.
22. Ibid., p.15.
23. Ibid., p. 11.
24. Historical Spotlight, 2001, ibid., p. 6.
25. Ibid., p. 28.
26. Ibid., p. 42.

27. Ibid., p. 62.
28. Historical Spotlight, 2000, ibid., pp. 20–55.
29. Barylski, ibid., pp. 299–333.
30. Tazpit, ibid., pp. 1–13.
31. Barylski, ibid., p. 314
32. Ibid., p. 382.
33. Historical Spotlight, 2001, ibid., pp. 22–23; Barylsk, ibid.
34. Historical Spotlight, ibid., pp. 37–38.
35. Historical Spotlight, 2000, ibid., pp.22.
36. Spotlight, 2000, ibid.; Spotlight 2001, ibid.

CHAPTER 6

1. Zigdon, Y. *Readings in Building of Force Theory* IDF Publication, 2004, (in Hebrew), pp. 259–269.
2. Ibid.
3. Ibid.
4. Ibid., p. 262
5. Ibid., p. 394
6. Cohen, S. A., & Bagno, O.-Y The societal consequences of military service in Israel: A

reappraisal. *Democratic Culture*, 4-5, 131–50 2001 (Hebrew).

7. Zigdon, ibid., p. 263.
8. Ofakim.org.il/zope/home/he/issue10/1098348124–38k.
9. Reported by Tsadok Yeheskeli, *Yediot A'haronot* 31.5.2002.
10. Ibid.
11. The discussion in this book does not deal with the role of the foreign media in its relationship with the army but focuses only on the internal media. It is important, however, to point out that, after the reports of the inquiries by the United Nations and of human rights organizations which rejected the claim of a massacre in Jenin, *Time* magazine published an investigation of its own which clearly rejected Palestinian claims and stated that the "*Time* investigation concludes that there was no wanton massacre in Jenin, no deliberate slaughter of Palestinians by Israeli soldiers."
12. Zigdon, ibid.
13. Zigdon, ibid.; Levy, Y. *The Other Army of Israel—Materialist Militarism in Israel* (Tel-Aviv, 2003) (Hebrew); Shelah, O. *The Tray and the Money* (Kinneret-Zmora Bitan, 2003) (in Hebrew).
14. Levi, 2003, pp. 418–419.
15. Levi, ibid., p. 239.
16. Zigdon, ibid., pp. 267–268.
17. Fishman Alex, *Yediot A'haronot* 10.3.2006.

18. Ibid.

19. Haaretz April 15, 2002.

20. Haaretz April 10, 2002.

21. Gideon Avidor, "The Battle of Jenin: April 2002," *The Urban Operation Journal*, http://www.urbanoperations.com/jenin.htm).

22. Guardian Unlimited, April 20, 2002, http://www.guardian.co.uk/international/story/0,3604,687644,00.html; Richard Hart, "A Hard Look at Jenin," *The Wall Street Journal*, April 30, 2002, p. A21.

23. http://www.shovrimshtika.org/press.asp.

24. http://www.shovrimshtika.org/what_we_did.asp.

25. http://www.tayasim.org.il.

26. http://www.seruv.org.il/hebrew/combatants_letter.asp.

27. http://my,ynet.co.il/pic/doc/211203/page1.htm.

28. Levy, 2003, p. 390.

29. Ibid., p. 418.

CONCLUSION

1. As stated, in every discussion of conscientious objection, reference is being made to refusal to be drafted to compulsory service for reasons of conscience. Reference is not being made to refusal to serve during the period of military service.

Bibliography

Alexander, M. S., Keiger J. F. V, Evans, M., (eds.). *The Algerian War and the French Army, 1954—62* (Palgrave: Macmillan, 2002) (translated to Hebrew).

Bartone, P. T., Adler, A. B., & Vaitkus, M. A. (1998). Dimensions of psychological stress in peacekeeping operations. *Military Medicine, 163*, 587–593.

Bartone, P. T. "Hardiness protects against war-related stress in Army reserve forces." *Consulting Psychology Journal* 51 (1999): 72–82.

Bartone, P. T. "Hardiness as a resiliency factor for United States forces in the Gulf War." In J. M. Violanti, D. Paton, & C. Dunning (eds.). *Posttraumatic Stress Intervention: Challenges, Issues, and Perspectives,* (Springfield, IL: Thomas, 2000), 115–133.

Bartone, P. T. "Resilience Under Military Operational Stress: Can Leaders Influence Hardiness?" *Military Psychology* (2006): 131–148.

Barylski, R. V. *The Soldier in Russian Politics (*New Brunswick: Transaction Publishers, 1998).

Bendor, G. "Military-Civil Relations in Israel in Mid 1990's" in Shapira A. (ed.), *Independence: Israel's Fifty Years* (Jerusalem: Zalman Shazar Press, 1998) (in Hebrew).

Ben Israel, I., "Security Theory and the Logic of Force Building," *Ma'arachot* (1997): 33–43 (in Hebrew).

Ben Israel, I., "Israel's Force Building Theory of Relativity" in *Ma'arachot* (1997): 252–352 (in Hebrew).

Bluth, C., *The Collapse of Soviet Military Power, (*Darmouth Publishing Company, England 1995).

Bowker, M. and Cameron R., (eds.). *Russia After the Cold War,* (Pearson Education Limited, England, 2000).

Britt, T. W. and Adler A. B. (eds.). *The Psychology of the Peacekeepers, (*Westport, CT: Praeger, 2003).

Brown, A., (ed.). *Contemporary Russian Politics, (*Oxford University Press, 2001).

Byrd, P. (ed.). *British Defence Policy,* (New York: P. Allan, 1991).

Calwell, C. E. *Small Wars and their Principles,* (Univ. of Nebraska Press, 1996).

Catterall, P. and McDougall, S. *The Northern Ireland Question in British Politics,* (Macmillan Press LTD, 1996).

Charters, D. and Tugwell, M. (eds.). *Armies in Low Intensity Conflict, (*Brassey's Defence Publishers, 1987).

Clinton, J. and Burke, M. D. "Doctrine for Asymmetric Warfare." *Military Doctrine,* July–August, (2003).

Cohen, A. *The Supreme Command—Military or Policy Makers: Who Leads in War?* (Ministry of Defence, 2003) (in Hebrew).

Cohen, S. (ed.) *Democratic Societies and their Armed Forces, (*London: BESA Studies in International Security, Frank Cass Publishers, 2000).

Dewar, M. *The British Army in Northern Ireland, (*Cassell Group Wellington House, 1985).

Dror, Y. *Grand Strategy for Israel,* (Leonard Davis Institute, Hebrew University 1999) (in Hebrew).

Dunlop, J. B. *Russia Confronts Chechnya: Roots of a Separatist Conflict,* Cambridge: (Cambridge University Press, 1998).

Figley, C. R., and Nash, W. P. (eds.) (2007). Combat Stress Injury: Theory, Research and Management, *Routledge Psychosocial Stress Series*, New York: Routledge.

Fishman, A. "Miluim Bemil." *Yediot Aharonot,* 3/10/2006 (in Hebrew).

Freeman, M. *Freedom or Security* (Westport: Praeger, 2003).

Gaidar, E. T. *Days of Defeat and Victory* (Moscow: Vagrius Publishers, 1996).

Gal, R. *A Portrait of the Israeli Soldier,* (Westport, CT: Greenwood Press, 1986).

Gal, R. "Military leadership for the 1990s: Commitment-derived leadership." In L. Atwater and R. Penn (eds.), *Military leadership: Traditions and future trends,* (Annapolis, MD: U.S. Naval Academy 1987: 53–59).

Gal, J. "What Really Erodes Society in a Long War?" *Mifne*, (January 2002) (in Hebrew).

Golan, H. and Shay, S. (eds.), *The Limited Conflict*, Tel-Aviv, 2004.

Hain, R.F. "The Use and Abuse of Technology in Insurgent Warfare." *Air & Space Chronicles,* (1999).

Hammes, T. X. *The Sling and the Stone: On War in the 21st Century,* (Zenith Press, 2004).

Harkabi, Y. *War and Strategy,* (Ministry of Defence, 1992) (in Hebrew).

Heggoy, A. A. *Insurgency and Counterinsurgency in Algeria,* (Indiana Univ. Press, 1972).

Henissart, P. *Wolves in the City,* (New York: Simon & Schuster Publishers, 1970).

Hennessey, T. *A History of Northern Ireland 1920–1996,* (St. Martin Press, 1997).

Horne, A. *A Savage War of Peace: Algeria 1954–1962,* (Viking Adult Publisher 1977) (translated to Hebrew).

Jacques D. and Ravid G, *Models of Foreign Armies*, A Sociel-Military project, IDF Manpower Division, 2001

Jauffret, J. *Soldats en Algerie 1954–1962, Experiences Contrastee des Hommes du contigent* (Paris : Editions Autrement, 2000).

Kearney, E. G. Creamer M., Marshall R. and Goyne A., (eds.). *Military Stress and Performance,* (Melbourne: Melbourne Univ. Press, 2003).

Kelly, K. K. *The Longest War: Northern Ireland and the IRA,* (UK: Zed Books, 1983).

Kennedy, P. *The Rise and Fall of the Great Powers,* (Davar Publications, 1992) (Translated to Hebrew).

Kobasa, S. "Commitment and Coping in Stress Resistance." *Journal of Personality and Social Psychology,* (1982) 42, 707–717.

Kober, A. "Low Intensity Conflict: Why the Gap Between Theory and Practice." *Defence and Security Analysis* (2002).

Kover, A. "Theory, Doctrines and Military Force Planning" *Ma'arachot,* (1985) (in Hebrew).

Kohn, R. K. "Out of Control: The Crisis in Civil Military Relations" *The National Interest,* No. 35 (Spring 1994).

Lazarus, R. S. *Psychological Stress and The Coping Process,* (New York: McGraw-Hill, 1996).

Levy, Y. The Other Army of Israel—Materialist Militarism in Israel, (Tel-Aviv, 2003) (Hebrew).

Lieven, A. *Tombstone of Russia Power,* (New Haven: Yale University Press, 1998).

Linn R. "Patterns of Crisis Among Israeli Reserve Soldiers."*The Jewish Journal of Sociology*, Vol. 39. Nos. 1 and 2, (1997): 24–45.

Majid, Al-Haj and Ben-Eliezer, U. *In the Name of Security,* (Haifa University Press/ Pardes Publishers 2003) (in Hebrew).

Maoz, Z. *Defending the Holy Land: a Critical Analysis of Israel's Security and Foreign Policy*, Ann Arbor: University Press (2006).

Massu, J. *La Vraie Bataille d'Alger,* (Paris: Plon, 1973).

Meichenbaum, D. and Jaremko M. E. *Stress Reduction and Prevention,* (Plenum Publishing Corporation, 1984).

Melman, Y. "Israeli Security Officials will Translate American Anxiety to Dollars" *Ha'Aretz,* (07/06/2003) (in Hebrew).

Moscos, C. "Towards a Postmodern Military?" In Cohen S. (ed.), *Democratic Societies and Their Armed Forces,* (Frank Cass, 2000).

Muelle, R. *La Guerre d Algerie en France,* Presses de la Cite, (1994).

Neuberger, B. (ed.). *Democracy and National Security,* (The Open University, 1996) (in Hebrew).

Posen, B. R. *The Sources of Military Doctrine: France, Britain, and Germany between the World Wars,* (Ithaca, NY and London: Cornell University Press, 1984).

Pye, L. "The Roots of Insurgency and the Commencement of Rebellions" in Eckstein, H. (ed.), *Internal War: Problems and Approaches.* (New York: Free Press of Glencoe (1964).

Rioux, P. and Sirinelli J. F. *La Guerre d'Algerie et les Intellectuels Francais,* (Bruxelles: Editions Complexe 1991).

Sarkesian, Sam C., and Scully, William L. (eds.). *U.S. Policy and Low-Intensity Conflict,* (New Brunswick: Transaction Books, 1981), pp. 20–21.

Schultz, G. "Low Intensity Warfare: The Challenge of Ambiguity" *U.S. Department of State Current Policy*, No.783, (Jan.1986).

Shelah, O. *The Tray and the Money,* (Kinneret-Zmora Bitan, 2003) (in Hebrew).
Shephard, B. *A War of Nerves,* (Harvard University Press, 2003).
Shrader, Charles R. *The First Helicopter War,* (Westport: Praeger, 1999).
Smith, S. *Politics and War in the Russian Caucasus,* (London: IB Tauris, 1998).
Strachan, H. *The Politics of the British Army,* (Oxford: Clarendon Press, 1997).
Stouffer, S.A. *The American Soldier,* (Princeton, NJ: Princeton University Press, 1949).
Talbott, J. *The War Without a Name: The French in Algeria 1954–1962,* (New York: Knopf, 1980).
Taylor, P. *Families at War,* (BBC Books, 1989).
Taylor, P. *Loyalists,* (Bloomsburg Pub., 1999).
Thackrah, R. J. *Dictionary of Terrorism,* (London: Routledge, 2004).
Tishkov, V. *The Making of a War Torn Society,* (University of California Press, 2004).
Tonge, J. *Northern Ireland—Conflict and Change,* (Prentice Hall Europe, 1998).
Tripier, P. *Autopsie de la Guerre d'Algerie,* (Editions France Empire, 1972).
Van Creveld, M. *Transformation of War,* (The Free Press, 1991).
Westman, M., "The Moderating Effect of Hardiness on the Relationship Between Stress and Performance." *Human Performance* 3, (1990: 141–155).
Whittaker, D. J. (ed.). *The Terrorism Reader,* (London: Routledge, 2003).
William, J.A. "The Post Modern Military Reconsidered," in Moskos, C., Segal et al (eds.). *The Postmodern Military* (NY: Oxford University Press, 2000).
Yaniv, A. (ed.).. *National Security and Democracy in Israel* (Boulder, CO: Lynne Rienner Publishers, 1993).
Zigdon, Y. *Readings in Building of Force Theory,* (IDF Publication, 2004: 259) (in Hebrew).

INTERNET AND RESEARCH PAPER SOURCES

Bell, M. "The Five Principles of Organizational Resilience," Gartner Archive, www.gartner.com.
"Algerian War Readings." www.usfca.edu/fac-staff/webberm/algeria.htm.
"Russian Newspapers on Russia and Chechnya War fmso.leavenworth.army.mil/documents/chechiw.htm.
"A History of the North of Ireland." www.historyworld.net/wrldhis/PlainTextHistories.asp?historyid=aa72.
www.etext.org/Politics/INAC/british.counterinsurgency.
cain.ulst.ac.uk/othelem/media/mccann72.htm—75k.
cain.ulst.ac.uk/othelem/media/mccann72.htm—75k.
www.geocities.com/Pentagon/6453/nireland.html—48k.
en.wikipedia.org/wiki/Algerian_War_of_Independenc.
www.usfca.edu/fac-staff/webberm/algeria.htm.
Zarkor, November 2000 (in Hebrew).
Zarkor, April 2001 (in Hebrew).
MAMDA papers (in Hebrew).
Tazpit, November 1996 (in Hebrew).

Index

About the Author

Dr. Rachel Suissa is an adjunct professor and a research fellow in the School of Political Sciences and in the Department of Geography at the University of Haifa, Israel. Her research interests encompass multidisciplinary security and emergency studies, as well as security and peace epistomology. Dr. Suissa is active in European Security Research agendas within the Seventh Framework Programme.

www.ingramcontent.com/pod-product-compliance
Ingram Content Group UK Ltd.
Pitfield, Milton Keynes, MK11 3LW, UK
UKHW012041150926
475560UK00001B/122

9 780739 12832